The Unknowable and the Counterintuitive

The Surprising Insights of Modern Science

Aleksandar I. Zecevic

University Readers™
San Diego, CA

Bassim Hamadeh, CEO and Publisher
Christopher Foster, General Vice President
Michael Simpson, Vice President of Acquisitions
Jessica Knott, Managing Editor
Stephen Milano, Creative Director
Kevin Fahey, Cognella Marketing Program Manager
Melissa Accornero, Acquisitions Editor

First published in the United States of America in 2012 by University Readers, Inc.

16 15 14 13 12 1 2 3 4 5

Printed in the United States of America

ISBN: 978-1-60927-494-8

University Readers™
800.200.3908 | www.universityreaders.com

Contents

Preface

This book grew out of a course on science and religion that I have taught at Santa Clara University for a number of years. This course (whose title is: "Chaos Theory, Metamathematics and the Limits of Knowledge: A Scientific Perspective on Religion") is rather unique, since it is *both* a technical elective for engineering students *and* satisfies a Core Curriculum requirement in religious studies. As such, it requires two very different textbooks - one that focuses on questions related to science, aesthetics and theology (*Truth, Beauty and the Limits of Knowledge*), and this book, which represents its "technical companion".

One of the most distinguishing features of this "companion" volume is the breadth of the topics that are covered. Indeed, one seldom (if ever) sees chaos theory, metamathematics, quantum mechanics and the theory of relativity united under a single title. My decision to adopt such an unorthodox approach was motivated in large part by the recognition that all four disciplines challenge the traditional Newtonian paradigm, and suggest that we must adopt a more sophisticated view of physical reality. This new outlook entails (among other things) a complete shift in the way we think, and brings into question some of our most deeply ingrained beliefs about nature.

Giving up on our "common sense" world view is, of course, a difficult thing to do, and most of us have no appetite for such profound conceptual changes. In that respect, classical physics has a decided advantage, since it is capable of reconciling theoretical results with our everyday experience. This is why we still teach it in schools, and why this approach continues to be popular with the general public. The fact remains, however, that the Newtonian methodology is limited to a fairly narrow range of "well behaved" phenomena, and that it fails to describe many subtle aspects of nature. Those who have some scientific training are well aware of this deficiency, but even they usually encounter it only within a particular area of specialization. In view of that, I decided to write a book that combines insights from several different disciplines, and outlines what each of them has to say about the counterintuitive nature of physical reality, and the limitations of human knowledge. This is the common theme that binds the many diverse topics that will be discussed in the following sections.

The book is divided into four chapters, each of which focuses on a different area of modern science and mathematics. The chapters are self-contained, and can be read without reference to any other parts of the book. Although such a structure provides readers with considerable flexibility (and even allows them to "pick and choose" according to their interests), I would advise against such a selective approach. I strongly believe that reading the entire text is a worthwhile investment, since only a broad interdisciplinary perspective can provide us with a proper understanding of the intricate laws that govern the world we live in. I would therefore encourage the reader to be curious, and explore as much "uncharted territory" as possible.

This brings me to my second reason for writing this book, which has to do with the relationship between science, philosophy and theology. Although most of the topics are highly technical, the overall text is designed to serve as a sort of "catalyst" for a more general interdisciplinary discussion. Indeed, those who manage to fully grasp some of the strange and elegant results of modern science often become more open to questions regarding the nature of reality. This is usually the point when scientists can begin to appreciate philosophical and theological insights, and can engage in a meaningful conversation with these disciplines. It is my hope that those who carefully read the following chapters will become amenable to such a possibility, and might even begin to see certain subtle similarities between the scientific and theological world views. If they do, they will join a very distinguished group of individuals, which includes Newton, Leibnitz, Einstein, Heisenberg, Gödel, Weyl and many other great names in the history of science and mathematics.

The material covered in this book is presented in a style that is typical for upper division textbooks in science and engineering. All important ideas and results are illustrated by examples, which should hopefully help the reader gain a better understanding of certain fundamental concepts in system theory, metamathematics and modern physics. Since chaos theory happens to be the primary technical topic in my course, I chose to devote more attention to this field than to others. As a result, my discussion of metamathematics, quantum mechanics and relativity is more selective, and is presented with less detail.

Chapter 1, which is concerned with chaos theory, opens with an overview of linear systems. This is an essential prerequisite for understanding the dynamics of nonlinear systems, which are considered in the following section. Our discussion of this topic will include methods for computing the system equilibria, an analysis of stability properties, and a description of different kinds of attractors. Some characteristic features of chaos (such as hypersensitivity to initial conditions and the fractal dimension of strange attractors) are introduced in Section 1.3, which is followed by a discussion of various types of bifurcations. Section 1.5 examines several different mechanisms that can lead to chaotic dynamics, and the chapter concludes with two case studies - the forced pendulum, and Chua's nonlinear circuit.

Chapter 2 introduces the field of metamathematics, and provides a brief description of formal systems. This material constitutes the necessary background for understanding Gödel's Incompleteness Theorem, whose proof is outlined in Section 2.3. The final section of this chapter describes some important consequences of Gödel's result, and examines several examples of incompleteness in modern mathematics.

Chapter 3 is devoted to quantum mechanics (in its non-relativistic form). The basic mathematical formalism that is used in this discipline is introduced in Section 3.1. We then proceed to discuss Heisenberg's Uncertainty Principle and the phenomenon of quantum entanglement, both of which impose fundamental limits on what we can know about the microscopic world. Section 3.4 also provides a brief description of Bell's inequality, and the role that it played in resolving the famous EPR paradox. The final section of this chapter introduces some basic concepts from group theory, which is one of the most important mathematical tools used in quantum mechanics and particle physics.

Chapter 4 consists of two separate parts, the first of which deals with special relativity. This section provides an introduction to relativistic mechanics, and describes some of its most important results (such as the Lorentz contraction, time dilation and Einstein's famous equation $E = mc^2$). In Section 4.2 we turn our attention to general relativity, whose main ideas are described in the framework of differential geometry. The discussion in this section focuses on the computation of geodesics, Christoffel symbols and the Schwarzschild metric.

Acknowledgements

My work on this book was supported by a grant from the Center for Science, Technology and Society at Santa Clara University. I am grateful to the Center for its generosity and its support for this project. I am also indebted to my colleagues Ruth Davis (Computer Engineering), Radovan Krtolica (Electrical Engineering) and Betty Young (Physics) who read the manuscript and provided many constructive suggestions.

Chapter 1

Chaos Theory

The phenomenon of chaos was first observed in 1961, by meteorologist Edward Lorenz. While studying a simple model for weather prediction, Lorenz noticed that repeated computer simulations produced very different results, although all the relevant system parameters appeared to be unchanged. After carefully examining all the possibilities, he ultimately established that the discrepancy was caused by minute rounding errors in the initial conditions (at the fifth decimal place). This was a thoroughly unexpected conclusion, since the underlying equations were deterministic, and should therefore be insensitive to such small variations. Lorenz subsequently named this phenomenon the "butterfly effect", a term that underscores the possibility that something as insignificant as the movement of a butterfly's wings could ultimately affect global weather patterns.

To understand why this result was so surprising, we should note that until the early 1960s physical processes were routinely classified as *either* deterministic *or* uncertain, but never both at once. The dynamic behavior of deterministic systems was assumed to be completely predictable, provided that all the parameters and initial conditions were known with sufficient accuracy. It was believed, in other words, that small changes in the initial conditions should *not* have a noticeable effect on the system trajectories. The model that Lorenz discovered clearly violated this premise, since it showed that precise mathematical models needn't always give rise to orderly behavior.

The realization that deterministic systems can sometimes exhibit unpredictable dynamics has important consequences for our understanding of science and its methods of investigation. To begin with, chaos theory shows that there are phenomena that do not lend themselves to repeated experimentation, since even the most minute variations in their surroundings can completely alter the outcome of the measurement. It goes with out saying, of course, that such variations are unavoidable, even in a strictly controlled environment. To this we should add the observation that the hypersensitivity of such systems to initial conditions makes it impossible to accurately simulate their long term behavior. Eliminating (or even minimizing) significant long term errors would require *unlimited precision*, which obviously cannot be accomplished in practice. It is therefore entirely appropriate to conclude that we can never have complete knowledge of such systems, regardless of how sophisticated our technology becomes.

Chaotic systems also challenge our intuitive notion of a physical "law", which implicitly assumes a certain level of regularity and predictability. To see why this is so, we should note that some systems of this type exhibit a property known as *intermittency*, which allows for sudden and completely unexpected aperiodic bursts that interrupt long periods of orderly behavior. Since these systems behave in the "usual" way both before and after such events, it is conceivable that entire generations of observers could register only the regular dynamic pattern, with no idea that other patterns are even possible. And yet, the underlying equations clearly allow for such anomalies, although they cannot predict when (or if) they will occur. Under such circumstances, it is quite possible that we may have to change our conventional understanding of physical laws, and reevaluate the extent to which they are "binding". Indeed, chaos theory suggests that laws governing complex systems may not be completely prescriptive, and can instead be viewed as a set of constraints, which determine what is and what is not possible. The details, however, will necessarily remain unknown to us.

Based on what we have learned about complex systems in recent decades, it seems that the laws governing their behavior are quite "liberal," and allow for an almost unlimited variety of configurations. This apparent "freedom" which is tempered by certain fairly moderate restrictions plays a crucial role in the process of *self-organization*. Current research suggests that complex systems (both of the animate and inanimate variety) seem to "prefer" operating in this mode, where small environmental changes can lead to the emergence of new and unprecedented natural forms. Behavior of this sort (which resembles the dynamics of systems that are on the "edge" of chaos) is seen by many as an essential prerequisite for the process of biological evolution.

In the sections that follow, we will examine some of these issues in greater detail. Such an analysis will require the use of nonlinear differential equations, as well as some elements of stability theory. In order to introduce the necessary concepts, we will begin with a brief overview of linear systems and their properties. We will then turn our attention to more sophisticated topics such as strange attractors, bifurcations and intermittency. Our discussion will conclude with two case studies, which illustrate how chaos arises in realistic physical systems.

1.1 Linear Systems: An Overview

The dynamic behavior of physical systems can often be described by a set of n first order differential equations of the form

$$\dot{x} = Ax + Bu \tag{1.1}$$

In (1.1),

$$x(t) = \begin{bmatrix} x_1 \\ \vdots \\ x_n \end{bmatrix} \tag{1.2}$$

represents the *state vector*,

$$u(t) = \begin{bmatrix} u_1 \\ \vdots \\ u_m \end{bmatrix} \tag{1.3}$$

denotes the *input*, and A and B are constant matrices of dimension $n \times n$ and $n \times m$, respectively. The following two examples illustrate how models of this type arise naturally in engineering problems.

Example 1.1. Let us consider a body with mass m which is connected to a spring and moves along a friction-producing surface. A schematic diagram of this system is provided in Fig. 1.1, where $F(t)$ represents an external force (which can be time-varying in general).

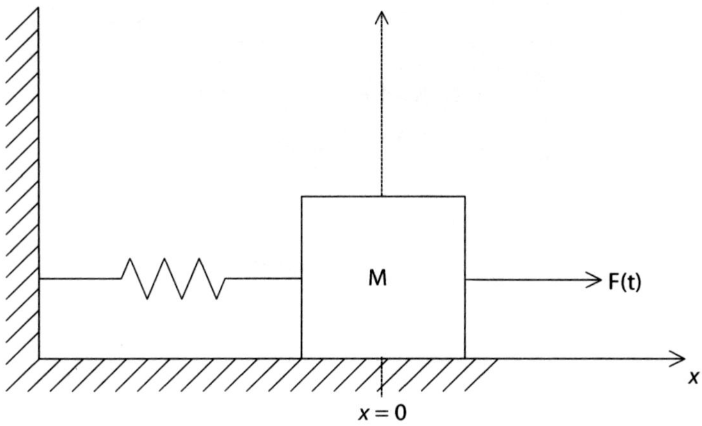

Fig. 1.1. A simple mechanical system.

It is normally assumed that friction is proportional to the velocity of the body, and that the restoring force of the spring has the form

$$F_s = -kx \tag{1.4}$$

In that case, Newton's Second Law produces the following equation of motion

$$m\ddot{x} = -\gamma\dot{x} - kx + F(t) \tag{1.5}$$

In order to obtain a state-space model, let us introduce state variables $x_1(t)$ and $x_2(t)$ as

$$\begin{aligned} x_1(t) &= x(t) \\ x_2(t) &= \dot{x}(t) \end{aligned} \tag{1.6}$$

Observing that $\ddot{x}(t) = \dot{x}_2(t)$, (1.5) can then be replaced by two first-order differential equations

$$\begin{aligned} \dot{x}_1 &= x_2 \\ \dot{x}_2 &= -(k/m)x_1 - (\gamma/m)x_2 + (1/m)F(t) \end{aligned} \tag{1.7}$$

This representation clearly conforms to (1.1), with

$$A = \begin{bmatrix} 0 & 1 \\ -k/m & -\gamma/m \end{bmatrix}, \qquad B = \begin{bmatrix} 0 \\ 1/m \end{bmatrix}, \qquad u(t) = F(t) \tag{1.8}$$

Example 1.2. Consider the circuit in Fig. 1.2, where $v_g(t)$ is assumed to be a time-varying voltage source.

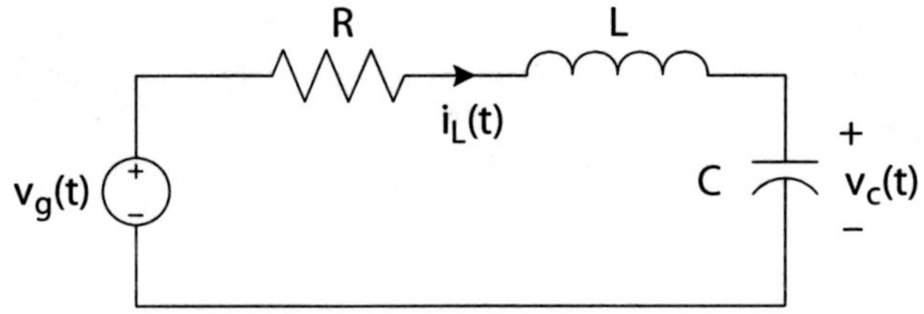

Fig. 1.2. A second order circuit.

Kirchoff's current and voltage laws for this circuit produce the following two equations

$$\begin{aligned} &i_C - i_L = 0 \\ &-v_g(t) + Ri_L + v_L + v_C = 0 \end{aligned} \tag{1.9}$$

Recalling that

$$v_L = L\frac{di_L}{dt} \quad \text{and} \quad i_C = C\frac{dv_C}{dt} \tag{1.10}$$

(1.9) becomes

$$\begin{aligned} C\dot{v}_C &= i_L \\ L\dot{i}_L &= -v_C - Ri_L + v_g(t) \end{aligned} \tag{1.11}$$

If we now define new variables $x_1(t)$ and $x_2(t)$ as

$$\begin{aligned} x_1(t) &= v_C(t) \\ x_2(t) &= i_L(t) \end{aligned} \tag{1.12}$$

we obtain a pair of first-order differential equations

$$\begin{aligned} \dot{x}_1 &= (1/C)x_2 \\ \dot{x}_2 &= -(1/L)x_1 - (R/L)x_2 + (1/L)v_g(t) \end{aligned} \tag{1.13}$$

This form obviously corresponds to (1.1), with

$$A = \begin{bmatrix} 0 & 1/C \\ -1/L & -R/L \end{bmatrix}, \qquad B = \begin{bmatrix} 0 \\ 1/L \end{bmatrix}, \qquad u(t) = v_g(t) \tag{1.14}$$

Eigenvalues and Eigenvectors

In the remainder of this section, we will consider systems of the form

$$\dot{x} = Ax \tag{1.15}$$

in which there are no external inputs (that is, $u = 0$). Since the dynamics of such systems depend primarily on A, the properties of this matrix need to be studied in some detail. We begin with the following definition.

Definition 1.1. A number λ_i (which can be real or complex) is said to be an *eigenvalue* of A if there is a nonzero vector y_i such that

$$Ay_i = \lambda_i y_i \tag{1.16}$$

A vector y_i that satisfies (1.16) is referred to as an *eigenvector* that corresponds to λ_i.

The computation of eigenvalues and eigenvectors is illustrated by the following example.

Example 1.3. Consider the linear system (1.15) where

$$A = \begin{bmatrix} 0 & 1 \\ -6 & -5 \end{bmatrix} \tag{1.17}$$

If λ is an eigenvalue of A, then equation

$$(A - \lambda I)y = 0 \tag{1.18}$$

must be satisfied for some nonzero vector y (here I denotes the identity matrix). Since system (1.18) has a nontrivial solution if and only if

$$\Delta(\lambda) = \det(A - \lambda I) = 0 \tag{1.19}$$

it follows that the eigenvalues of A are actually the roots of polynomial $\Delta(\lambda)$. In our example, this polynomial has the form

$$\Delta(\lambda) = \lambda^2 + 5\lambda + 6 = (\lambda + 2)(\lambda + 3) \tag{1.20}$$

which implies that the eigenvalues are $\lambda_1 = -2$ and $\lambda_2 = -3$.

How can we determine the eigenvectors that correspond to λ_1 and λ_2? For $\lambda_1 = -2$, (1.18) becomes

$$\begin{bmatrix} 2 & 1 \\ -6 & -3 \end{bmatrix} \begin{bmatrix} y_{11} \\ y_{12} \end{bmatrix} = \begin{bmatrix} 0 \\ 0 \end{bmatrix} \tag{1.21}$$

which is a singular system of equations. This property becomes obvious after applying Gaussian elimination, which produces

$$\begin{bmatrix} 2 & 1 \\ 0 & 0 \end{bmatrix} \begin{bmatrix} y_{11} \\ y_{12} \end{bmatrix} = \begin{bmatrix} 0 \\ 0 \end{bmatrix} \tag{1.22}$$

Observing that the second equation is satisfied for *any* choice of y_{11} and y_{12}, we can set $y_{12} = t$ (where t is an arbitrary real number). Component y_{11} can then be uniquely determined as

$$y_{11} = -\frac{t}{2} \tag{1.23}$$

The solution of (1.22) has the general form

$$y_1 = t \begin{bmatrix} -1/2 \\ 1 \end{bmatrix} \tag{1.24}$$

which suggests that there are actually *infinitely many* eigenvectors corresponding to λ_1. However, all of these eigenvectors are proportional to vector

$$\bar{y}_1 = \begin{bmatrix} -1/2 \\ 1 \end{bmatrix} \tag{1.25}$$

which we will choose as the 'representative' for this set.

We can proceed in a similar way to obtain the eigenvector corresponding to $\lambda_2 = -3$. In that case we have

$$\begin{bmatrix} 3 & 1 \\ -6 & -2 \end{bmatrix} \begin{bmatrix} y_{21} \\ y_{22} \end{bmatrix} = \begin{bmatrix} 0 \\ 0 \end{bmatrix} \tag{1.26}$$

which becomes

$$\begin{bmatrix} 3 & 1 \\ 0 & 0 \end{bmatrix} \begin{bmatrix} y_{21} \\ y_{22} \end{bmatrix} = \begin{bmatrix} 0 \\ 0 \end{bmatrix} \tag{1.27}$$

after Gaussian elimination. It is not difficult to verify that all the solutions of (1.27) have the form

$$y_2 = t \begin{bmatrix} -1/3 \\ 1 \end{bmatrix} \tag{1.28}$$

and that

$$\bar{y}_2 = \begin{bmatrix} -1/3 \\ 1 \end{bmatrix} \tag{1.29}$$

can be taken as the representative eigenvector.

Linear Transformations and Jordan Canonical Forms

A linear transformation represents a change of the coordinate system in which we view our model. To see what that means, let us introduce a new state vector

$$z(t) = T^{-1}x(t) \tag{1.30}$$

where T is a given *transformation matrix*. Since

$$x(t) = Tz(t) \tag{1.31}$$

the original system (1.1) can now be expressed as

$$T\dot{z} = ATz + Bu \tag{1.32}$$

Multiplying both sides on the left by T^{-1}, we obtain

$$\dot{z} = \tilde{A}z + \tilde{B}u \tag{1.33}$$

where

$$\tilde{A} = T^{-1}AT \quad \text{and} \quad \tilde{B} = T^{-1}B. \tag{1.34}$$

A particularly useful coordinate transformation corresponds to the case when T is formed from the eigenvectors of matrix A. To see why this is so, consider a 2×2 matrix A with a pair of real, distinct eigenvalues λ_1 and λ_2. If y_1 and y_2 are the associated eigenvectors, we have

$$\begin{aligned} Ay_1 &= \lambda_1 y_1 \\ Ay_2 &= \lambda_2 y_2 \end{aligned} \tag{1.35}$$

by definition. Forming matrix T as

$$T = \begin{bmatrix} y_1 & y_2 \end{bmatrix} \tag{1.36}$$

we obtain

$$AT = \begin{bmatrix} Ay_1 & Ay_2 \end{bmatrix} = \begin{bmatrix} \lambda_1 y_1 & \lambda_2 y_2 \end{bmatrix} = T\Lambda \tag{1.37}$$

where

$$\Lambda = \begin{bmatrix} \lambda_1 & 0 \\ 0 & \lambda_2 \end{bmatrix} \tag{1.38}$$

Since

$$\tilde{A} = T^{-1}AT = \Lambda \tag{1.39}$$

it follows that this transformation *diagonalizes* A. The elements of Λ are the eigenvalues of A, and this matrix is referred to as the *Jordan canonical form* of A.

Remark 1.1. When the eigenvalues are complex or multiple, the Jordan canonical form becomes more complicated. It should be noted, however, that the main ideas developed in the following are easily extended to this case.

The Jordan canonical form can greatly simplify the solution of the state equations. Indeed, suppose that we are given a system like (1.15) where A is an $n \times n$ matrix with real, distinct eigenvalues. If we transform this system into the Jordan form

$$\dot{z} = \Lambda z \tag{1.40}$$

it is easily verified that (1.40) now consists of n decoupled scalar equations

$$\begin{aligned} \dot{z}_1 &= \lambda_1 z_1 \\ &\vdots \\ \dot{z}_n &= \lambda_n z_n \end{aligned} \tag{1.41}$$

whose solutions are

$$\begin{aligned} z_1(t) &= z_1(0)\, e^{\lambda_1 t} \\ &\vdots \\ z_n(t) &= z_n(0)\, e^{\lambda_n t} \end{aligned} \tag{1.42}$$

The overall solution vector can be represented in compact form as

$$z(t) = \begin{bmatrix} e^{\lambda_1 t} & & \\ & \ddots & \\ & & e^{\lambda_n t} \end{bmatrix} z(0) \tag{1.43}$$

where $z(0) = T^{-1}x(0)$.

To see how (1.43) appears in the original coordinates, we should recall that

$$x(t) = Tz(t) = \begin{bmatrix} y_1 & \dots & y_n \end{bmatrix} \begin{bmatrix} e^{\lambda_1 t} & & \\ & \ddots & \\ & & e^{\lambda_n t} \end{bmatrix} z(0) \tag{1.44}$$

which becomes

$$x(t) = z_1(0)\, e^{\lambda_1 t} y_1 + \ldots + z_n(0)\, e^{\lambda_n t} y_n \tag{1.45}$$

after multiplication. A solution expressed in this form is often referred to as the *modal decomposition*. It has the advantage of explicitly showing how the system dynamics depend on the eigenvalues and eigenvectors of A.

Example 1.4. Let us determine the modal decomposition of system (1.15), whose eigenvalues and eigenvectors were computed in Example 1.3. Using (1.25) and (1.29), we directly obtain

$$x(t) = z_1(0)\, e^{-2t} \begin{bmatrix} -1/2 \\ 1 \end{bmatrix} + z_2(0)\, e^{-3t} \begin{bmatrix} -1/3 \\ 1 \end{bmatrix} \tag{1.46}$$

The initial conditions $z_1(0)$ and $z_2(0)$ can be related to $x(0)$ by observing that $z(0) = T^{-1}x(0)$, where

$$T = \begin{bmatrix} -1/2 & -1/3 \\ 1 & 1 \end{bmatrix} \tag{1.47}$$

Since

$$T^{-1} = \begin{bmatrix} -6 & -2 \\ 6 & 3 \end{bmatrix} \tag{1.48}$$

it follows that

$$\begin{aligned} z_1(0) &= -6x_1(0) - 2x_2(0) \\ z_2(0) &= 6x_1(0) + 3x_2(0) \end{aligned} \tag{1.49}$$

Equilibria and Stability

Any constant solution of (1.15) is referred to as an *equilibrium* of the system. Since such a solution must satisfy

$$Ax^e = 0 \tag{1.50}$$

by definition, it follows that $x^e = 0$ is a *unique* equilibrium whenever A is a nonsingular matrix.

Definition 1.2. Equilibrium x^e is a *global attractor* if the solution satisfies

$$x(t) \to x^e \qquad (t \to \infty) \tag{1.51}$$

for *any* initial condition $x(0)$. Such an equilibrium is also said to be *globally asymptotically stable.*

Attractivity is a very important characteristic of any equilibrium. Recalling (1.45), we can conclude that $x^e = 0$ will have this property when all eigenvalues satisfy $\lambda_i < 0$. It can be shown that a similar condition holds even if the eigenvalues are complex, in which case the requirement becomes $\text{Re}(\lambda_i) < 0$, $(i = 1, 2, \ldots, n)$.

If one or more eigenvalues satisfy $\text{Re}(\lambda_i) > 0$, $x^e = 0$ does not attract the solutions any more. In that case, the equilibrium is said to be *unstable* (it can also be referred to as a *repeller*). The following example illustrates such a situation.

Example 1.5. Consider a system of the form (1.15), in which

$$A = \begin{bmatrix} 0.5 & 1.5 \\ 1.5 & 0.5 \end{bmatrix} \tag{1.52}$$

The eigenvalues of A are $\lambda_1 = -1$ and $\lambda_2 = 2$, and the corresponding eigenvectors are

$$y_1 = \begin{bmatrix} -1 \\ 1 \end{bmatrix} \quad \text{and} \quad y_2 = \begin{bmatrix} 1 \\ 1 \end{bmatrix}. \tag{1.53}$$

Introducing $z(t) = T^{-1}x(t)$ where

$$T = \begin{bmatrix} -1 & 1 \\ 1 & 1 \end{bmatrix} \tag{1.54}$$

we obtain two decoupled equations

$$\begin{aligned} \dot{z}_1 &= -z_1 \\ \dot{z}_2 &= 2z_2 \end{aligned} \tag{1.55}$$

with solutions

$$\begin{aligned} z_1(t) &= z_1(0)\, e^{-t} \\ z_2(t) &= z_2(0)\, e^{2t} \end{aligned} \tag{1.56}$$

The modal decomposition will therefore have the form

$$x(t) = z_1(0)\, e^{-t} \begin{bmatrix} -1 \\ 1 \end{bmatrix} + z_2(0)\, e^{2t} \begin{bmatrix} 1 \\ 1 \end{bmatrix} \tag{1.57}$$

Since one of the exponents is positive, we can conclude that $x(t) \to \infty$ $(t \to \infty)$ for almost all initial conditions. What do we mean by 'almost all'? To see this, it suffices to observe that there are some special initial conditions for which $x(t)$ still tends to $x^e = 0$ when $t \to \infty$. These initial conditions correspond to $z_2(0) = 0$, since in that case

$$x(t) = z_1(0)\, e^{-t} \begin{bmatrix} -1 \\ 1 \end{bmatrix} \tag{1.58}$$

We can identify the set of all such initial conditions explicitly, by recalling that

$$\begin{bmatrix} z_1(0) \\ z_2(0) \end{bmatrix} = T^{-1} \begin{bmatrix} x_1(0) \\ x_2(0) \end{bmatrix} = \begin{bmatrix} -0.5 & 0.5 \\ 0.5 & 0.5 \end{bmatrix} \begin{bmatrix} x_1(0) \\ x_2(0) \end{bmatrix} \tag{1.59}$$

The requirement then becomes

$$z_2(0) = 0.5x_1(0) + 0.5x_2(0) = 0 \tag{1.60}$$

which means that $x(t) \to 0$ if and only if

$$x_1(0) = -x_2(0). \tag{1.61}$$

Condition (1.61) defines a line in the $x_1 - x_2$ space which is known as a *stable manifold*. All solutions starting on the manifold approach the equilibrium when $t \to \infty$, while solutions originating outside it increase unboundedly.

1.2 Nonlinear Systems

A nonlinear dynamic system is generally described by a set of first order differential equations

$$\dot{x} = f(x, t) \tag{1.62}$$

where $f(x,t)$ is a vector-valued nonlinear function. Models of this type arise in a wide variety of applications, ranging from engineering to economics and ecology. From a theoretical standpoint, it is important to distinguish between *autonomous* and *non-autonomous* dynamic systems. An example of a non-autonomous system would be

$$\begin{aligned} \dot{x}_1 &= -x_1 + 2x_1x_2 + x_2^2 \\ \dot{x}_2 &= x_1 + x_2 + x_1^2x_2 + 2\cos 3t \end{aligned} \tag{1.63}$$

where time shows up *explicitly* through the cosine term. In such cases it is appropriate to use the general description (1.62), since f is a function of both x

and t. In contrast, an autonomous system has no explicit time dependency, and can be described simply as

$$\dot{x} = f(x) \tag{1.64}$$

The second order system

$$\begin{aligned} \dot{x}_1 &= x_1 - x_2^2 \\ \dot{x}_2 &= x_1 \sin x_2 - x_2 \end{aligned} \tag{1.65}$$

is a typical example.

To see how nonlinear systems differ from their linear counterparts, let us consider a simple scalar system

$$\dot{x} = x^2 + 2x \tag{1.66}$$

with a quadratic nonlinearity. The first thing that we notice about it is that it has *multiple equilibria*, corresponding to $x^e = 0$ and $x^e = -2$. This is quite unlike linear systems, where the coordinate origin was always the unique equilibrium, regardless of the size of the model (assuming, of course, that matrix A is nonsingular). In dealing with multiple equilibria, two questions arise naturally:

1. How can we compute all the equilibria of a general dynamic system?

2. Is it possible to determine whether a particular equilibrium is an attractor or a repeller?

In the following, we will address these issues in some detail.

Computation of Equilibria

The system equilibria can be viewed as solutions of the algebraic equation

$$f(x) = 0 \tag{1.67}$$

Computing the solutions of (1.67) can be a very difficult task in general, and there are no guarantees that all of them will be found. One of the most effective solution techniques for nonlinear equations is the so-called *Newton iterative method*. The following simple example illustrates the basic ideas of this approach.

Example 1.6. Consider the scalar equation

$$f(x) = x^2 + 4x + 3 = 0 \tag{1.68}$$

which we propose to solve iteratively. In order to do that, we will form the following sequence

$$x(k+1) = x(k) - \left[f'(x(k))\right]^{-1} f(x(k)) \tag{1.69}$$

$x(0) = 0$	$x(0) = -5$
$x(1) = -0.7500$	$x(1) = -3.6700$
$x(2) = -0.9750$	$x(2) = -3.1330$
$x(3) = -0.9969$	$x(3) = -3.0078$
$x(4) = -0.9999$	$x(4) = -3.0000$

Table 1.1. Iterative sequences for two different initial guesses.

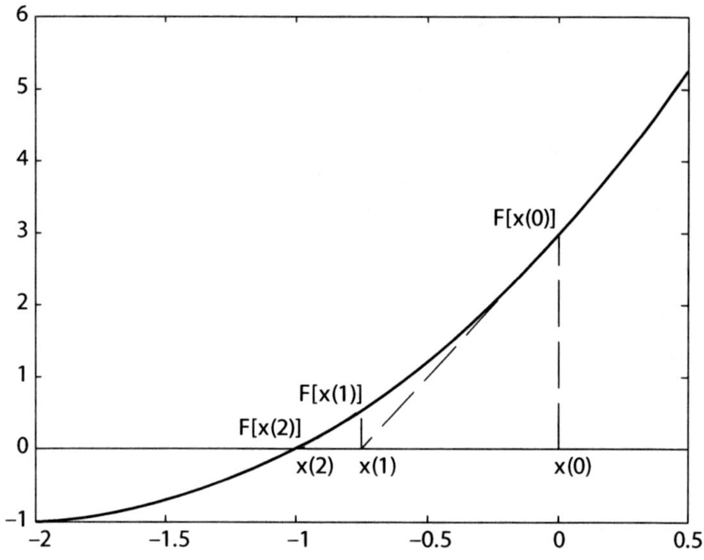

Fig. 1.3. An illustration of Newton's method.

starting from some initial guess $x(0)$. Two possible scenarios are shown in Table 1.1, suggesting that different choices of $x(0)$ can lead to very different solutions.

Before extending Newton's method to larger systems, it is useful to consider its geometric interpretation in the scalar case. To that effect, in Fig. 1.3 we show the plot of function (1.68) in a region that contains the solution $x^* = -1$.

The point labeled $x(1)$ represents the intersection of slope $f'(x(0))$ with the x-axis. As a result, we can write

$$f'(x(0)) = \frac{f(x(0))}{x(0) - x(1)} \tag{1.70}$$

which directly implies

$$x(1) = x(0) - [f'(x(0))]^{-1} f(x(0)) \tag{1.71}$$

Equation (1.71) allows us to conclude that point $x(1)$ on the graph corresponds to the first iterate obtained using Newton's method. It is not difficult to see that

$x(2)$, $x(3)$, etc. have the same interpretation, and that this sequence converges to $x^* = -1$.

In the multivariable case, Newton's method takes the form

$$x(k+1) = x(k) - [J(x(k))]^{-1} f(x(k)) \tag{1.72}$$

where

$$f(x) = \begin{bmatrix} f_1(x_1, x_2, \dots, x_n) \\ \vdots \\ f_n(x_1, x_2, \dots, x_n) \end{bmatrix} \tag{1.73}$$

is a vector-valued nonlinear function, and

$$J(x) = \begin{bmatrix} \partial f_1/\partial x_1 & \cdots & \partial f_1/\partial x_n \\ \vdots & \ddots & \vdots \\ \partial f_n/\partial x_1 & \cdots & \partial f_n/\partial x_n \end{bmatrix} \tag{1.74}$$

represents its Jacobian. The following example illustrates how the iterative process works in this case.

Example 1.7. Consider the pair of nonlinear equations

$$f(x) = \begin{bmatrix} 3x_1x_2 - x_2^{-2} - e^{x_1} \\ x_1^2 x_2 - \cos x_1 - x_2 \end{bmatrix} = 0 \tag{1.75}$$

whose solutions we wish to compute. It is easily verified that the Jacobian of this system has the form

$$J(x) = \begin{bmatrix} 3x_2 - e^{x_1} & 3x_1 + 2x_2^{-3} \\ 2x_1x_2 + \sin x_1 & x_1^2 - 1 \end{bmatrix} \tag{1.76}$$

In Table 1.2, we show Newton's iterates for two different initial guesses.

$x(0) = [1 \quad 1]$	$x(0) = [-1 \quad -1]$
$x(1) = [1.1901 \quad 1.1329]$	$x(1) = [-0.5336 \quad -0.9877]$
$x(2) = [1.1619 \quad 1.1379]$	$x(2) = [-0.4134 \quad -1.1120]$
$x(3) = [1.1617 \quad 1.1383]$	$x(3) = [-0.4330 \quad -1.1167]$
$x(4) = [1.1617 \quad 1.1383]$	$x(4) = [-0.4328 \quad -1.1170]$

Table 1.2. Iterative sequences for two different initial guesses.

As in Example 1.6, it is apparent that different initial guesses produce different solutions of (1.75). In this case, however, it is very difficult to tell whether additional solutions exist (and what initial guesses would lead to them). The same can be said for a broad class of similar problems where analytical solutions are not available.

Remark 1.2. In general, there is no guarantee that Newton's method will converge. Whether or not that will happen depends to a large extent on the choice of initial approximation. This can be a major problem, particularly when the system is large.

Stability

A simple way to determine whether a given equilibrium x^e is attractive or not is to *linearize* the system using the Taylor series expansion. For the system shown in (1.66), linearization around $x^e = 0$ yields

$$\dot{x} = 2x \tag{1.77}$$

which indicates that this equilibrium is a *repeller*. Similarly, linearization around $x^e = -2$ yields

$$\dot{y} = -2y \tag{1.78}$$

where $y = x - x^e$. This implies that $x^e = -2$ is an *attractor*.

It is important to recognize that linearization provides only *local* information about an equilibrium. As a result, we can claim that trajectories originating close to $x^e = -2$ will be attracted by this equilibrium, but we can't really say anything about other solutions of the system. This is another fundamental difference with respect to linear systems, where an asymptotically stable equilibrium was a *global attractor*.

To further illustrate the local properties of equilibria, let us consider the system

$$\begin{aligned} \dot{x}_1 &= -3x_1 + 4x_1^2 - 0.5x_1x_2 - x_1^3 \\ \dot{x}_2 &= -2.1x_2 + x_1x_2 \end{aligned} \tag{1.79}$$

which represents a variant of the classical predator-prey model in ecology (x_1 and x_2 are the populations of the prey and predator, respectively).

A. The Equilibria

Our first step will be to determine the equilibria of this system, which implies solving the following pair of equations

$$\begin{aligned} -3x_1 + 4x_1^2 - 0.5x_1x_2 - x_1^3 = 0 \\ -2.1x_2 + x_1x_2 = 0 \end{aligned} \tag{1.80}$$

In a typical nonlinear system we would have to do this numerically, with no guarantee of finding all possible solutions. In this case, however, the problem can be solved analytically, by observing that condition

$$x_2(x_1 - 2.1) = 0 \tag{1.81}$$

allows for two possible cases:

CASE 1. $x_2 \neq 0$, which implies that $x_1 = 2.1$ and $x_2 = 1.98$ is the unique solution.

CASE 2. $x_2 = 0$, which implies that x_1 can be determined by solving equation

$$-3x_1 + 4x_1^2 - x_1^3 = 0 \tag{1.82}$$

This is a simple third order polynomial with roots 0, 1 and 3.

Based on the above analysis, we can conclude that the predator-prey model has *exactly* four equilibria

$$x^e = \begin{bmatrix} 0 \\ 0 \end{bmatrix}; \quad x^e = \begin{bmatrix} 1 \\ 0 \end{bmatrix}; \quad x^e = \begin{bmatrix} 3 \\ 0 \end{bmatrix}; \quad x^e = \begin{bmatrix} 2.10 \\ 1.98 \end{bmatrix} \tag{1.83}$$

B. Stability

Turning now to the question of stability, we first need to linearize the system around each equilibrium. The Jacobian of (1.80) has the general form

$$J(x) = \begin{bmatrix} (-3 + 8x_1 - 0.5x_2 - 3x_1^2) & -0.5x_1 \\ x_2 & -2.1 + x_1 \end{bmatrix} \tag{1.84}$$

which leads to the following scenarios:

(*i*) For $x^e = [0 \ \ 0]^T$, we have

$$J(x^e) = \begin{bmatrix} -3 & 0 \\ 0 & -2.1 \end{bmatrix} \tag{1.85}$$

The linearized system is therefore

$$\dot{y} = J(x^e)y \tag{1.86}$$

where $y = x - x^e$. Since this is now a system of the form (1.15), its stability properties can be determined by checking the eigenvalues. In this case, $\lambda_1 = -3$ and $\lambda_2 = -2.1$, indicating that $x^e = [0 \ \ 0]^T$ is a *stable* equilibrium.

(*ii*) For $x^e = [1 \ \ 0]^T$, we have

$$J(x^e) = \begin{bmatrix} 2 & -0.5 \\ 0 & -1.1 \end{bmatrix} \tag{1.87}$$

and the eigenvalues of the linearized system are $\lambda_1 = 2$ and $\lambda_2 = -1.1$. This indicates that $x^e = [1 \ \ 0]^T$ is an *unstable* equilibrium.

(*iii*) For $x^e = [3 \ \ 0]^T$, we have

$$J(x^e) = \begin{bmatrix} -6 & -1.5 \\ 0 & 0.9 \end{bmatrix} \tag{1.88}$$

and the eigenvalues of the linearized system are $\lambda_1 = -6$ and $\lambda_2 = 0.9$. This indicates that $x^e = [3 \ \ 0]^T$ is an *unstable* equilibrium.

(*iv*) For $x^e = [2.1 \quad 1.98]^T$, we have

$$J(x^e) = \begin{bmatrix} -0.42 & -1.05 \\ 1.98 & 0 \end{bmatrix} \tag{1.89}$$

and the eigenvalues of the linearized system are $\lambda_{1,2} = -0.21 \pm j1.43$. This indicates that $x^e = [2.1 \quad 1.98]^T$ is a *stable* equilibrium.

At this point, we know that two of the four equilibria are locally attractive. However, we still don't know what happens to solutions that don't originate in the immediate neighborhood of these equilibria. This is actually a very hard question to answer analytically, even for relatively simple systems. One way to approach the problem is to perform multiple simulations, and then plot each trajectory in the $x_1 - x_2$ plane. A collection of these trajectories is referred to as the *phase portrait* of the system, and is shown in Fig. 1.4 (the solid dots indicate the initial conditions).

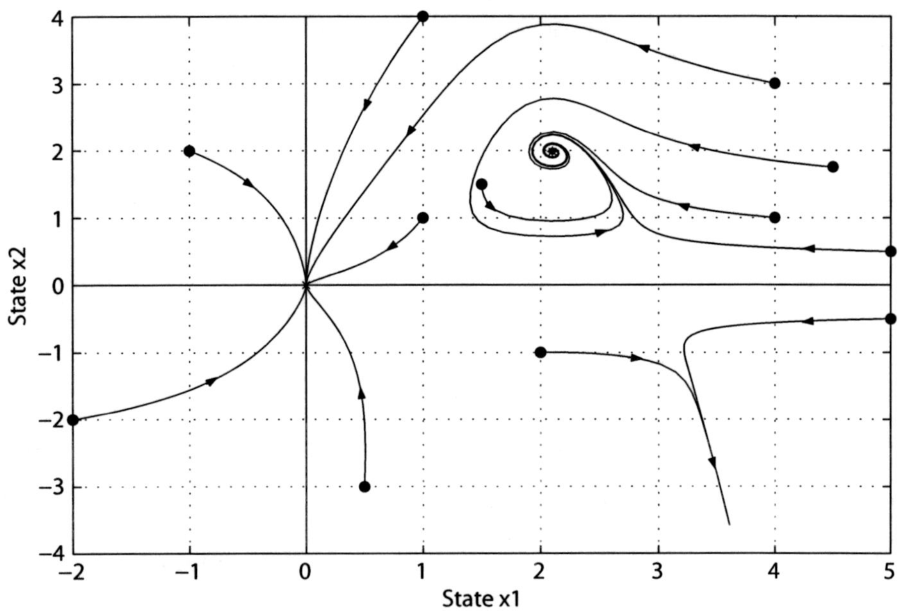

Fig. 1.4. Regions of attraction for different equilibria.

Fig. 1.4 suggests that there are entire regions in the phase space from which solutions are attracted to particular equilibria. Such regions are known as *regions of attraction*. In this context we should note that there may also be regions from which solutions are not attracted to any of the equilibria (an example of such a region is the lower right-hand quadrant of Fig. 1.4).

Types of Attractors in Nonlinear Systems

In all the examples that we considered so far, the trajectories were attracted to (or repelled by) the system equilibria. It turns out, however, that in some nonlinear systems trajectories can also be attracted by other objects in the phase space. In the following, we will identify three such objects, in increasing degree of complexity.

Limit Cycles

Let us consider the following system (also known as the van der Pol oscillator)

$$\begin{aligned} \dot{x}_1 &= x_2 \\ \dot{x}_2 &= (1 - x_1^2)x_2 - x_1 \end{aligned} \tag{1.90}$$

which has a unique equilibrium at the origin. Observing that the Jacobian has the form

$$J(x) = \begin{bmatrix} 0 & 1 \\ -(2x_1x_2 + 1) & 1 - x_1^2 \end{bmatrix} \tag{1.91}$$

it follows that

$$J(x^e) = \begin{bmatrix} 0 & 1 \\ -1 & 1 \end{bmatrix} \tag{1.92}$$

with eigenvalues $\lambda_{1,2} = 0.5 \pm j0.866$. This indicates that $x^e = [0 \;\; 0]^T$ is a *repeller*, and that solutions originating in the neighborhood of this point diverge from it. It is important to recall, however, that this linearization is valid only near the origin, so we can't tell what happens to these solutions as $t \to \infty$. The easiest way to evaluate the long term behavior of the trajectories is to numerically solve the differential equation for several different initial conditions (some near the equilibrium, and some further away). Typical responses for $x_1(t)$ and $x_2(t)$ are shown in Figs. 1.5 and 1.6, indicating that solutions seem to settle into a periodic steady state as $t \to \infty$. Such a trajectory is referred to as a *limit cycle*, and it represents a legitimate attractor for this system.

To see the attractive nature of the limit cycle more clearly, in Fig. 1.7 we provide a phase portrait of the system (making sure that the simulation time was long enough to allow solutions to reach their steady state). This portrait indicates that the limit cycle is represented by a *closed curve* in the phase space, and that all of our trajectories are attracted to it.

Quasi-Periodic Attractors

Before we begin our discussion of quasi-periodic attractors, it is first necessary to define what we mean by quasi-periodic behavior. The simplest example would be the function

$$f(t) = \cos \omega_1 t + \cos \omega_2 t \tag{1.93}$$

which is a sum of two sinusoids with different frequencies. Is such a function periodic? Well, that depends... The necessary condition for periodicity is that

$$f(t + T) = f(t) \tag{1.94}$$

holds for some T. It is easily verified that this will be the case when

$$\begin{aligned} \omega_1 T &= 2p\pi \\ \omega_2 T &= 2q\pi \end{aligned} \tag{1.95}$$

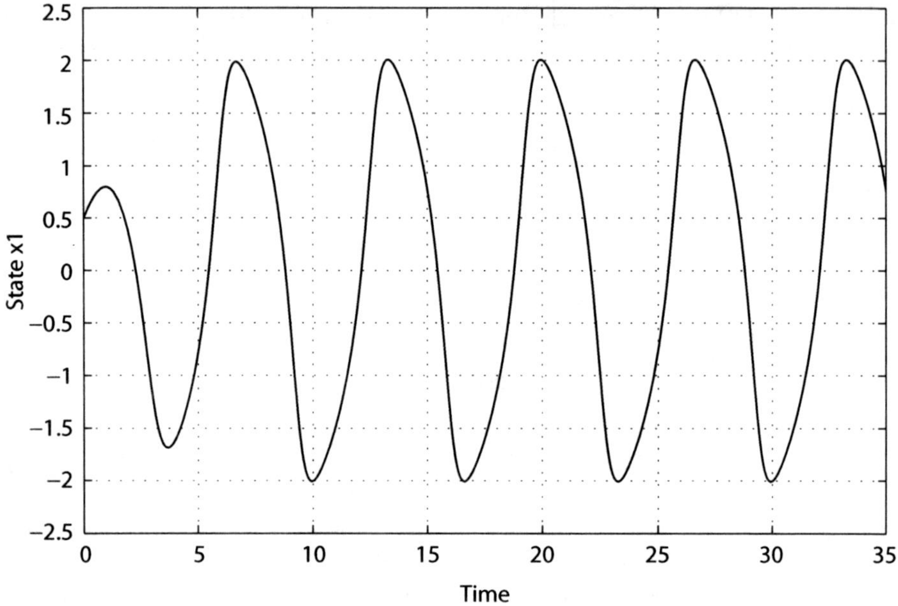

Fig. 1.5. Evolution of state x_1.

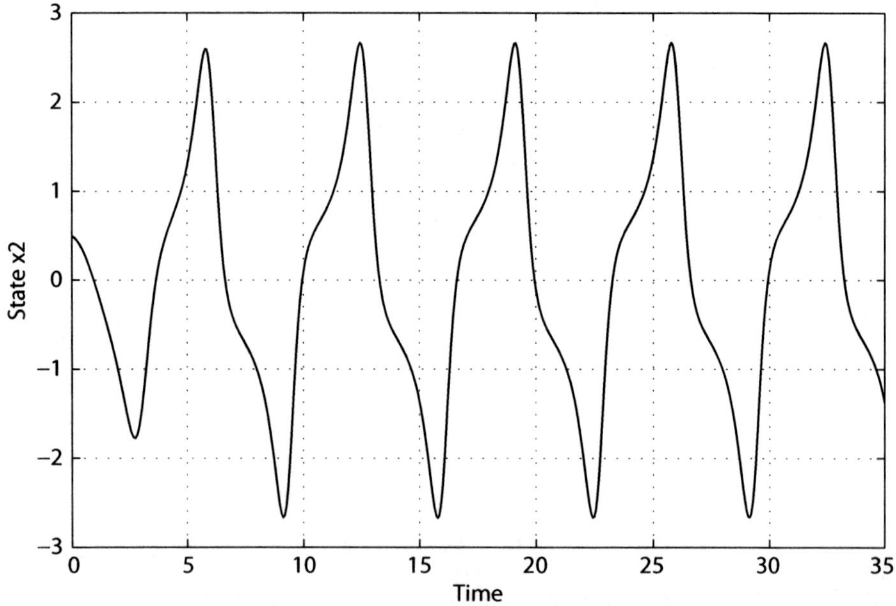

Fig. 1.6. Evolution of state x_2.

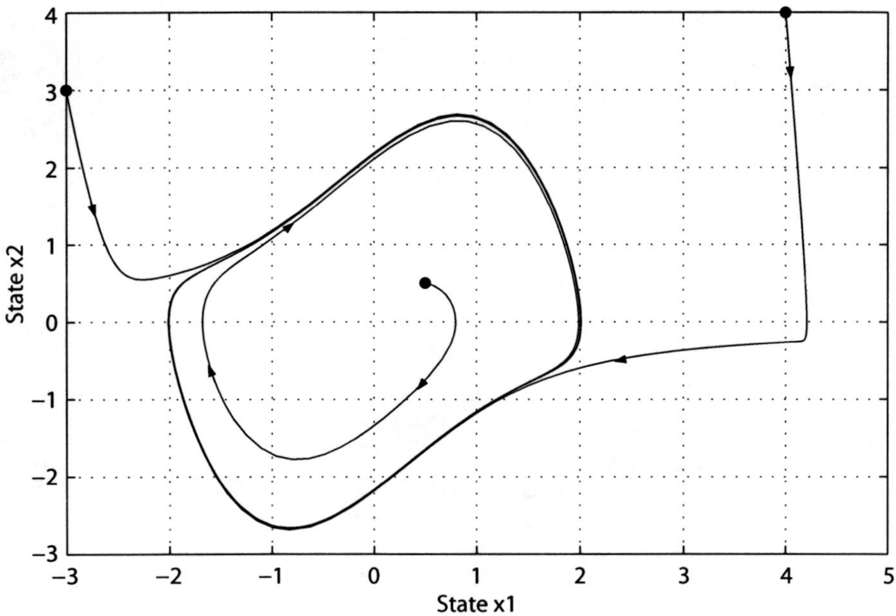

Fig. 1.7. The limit cycle in the phase space.

for some integers p and q, or equivalently if

$$\frac{\omega_1}{\omega_2} = \frac{p}{q} \tag{1.96}$$

When (1.96) holds, we say that frequencies ω_1 and ω_2 are *commensurate*, and $f(t)$ is periodic with

$$\omega_{eq} = \frac{2\pi}{T} = \frac{\omega_1}{p} = \frac{\omega_2}{q} \tag{1.97}$$

This scenario is often referred as *frequency locking*, since ω_{eq} 'locks' onto some fraction of the original frequencies ω_1 and ω_2.

What happens if ω_1 and ω_2 are *not* commensurate (for instance, when $\omega_1 = \sqrt{3}\,\omega_2$)? In that case, $f(t)$ is clearly not periodic, but it can nevertheless be represented as a sum of two independent periodic functions. Since $f(t)$ still exhibits some regularity in its behavior, we say that it is *quasi-periodic.*

The following simple example provides an illustration of both frequency locking and quasi-periodic behavior in a physical system.

Example 1.8. The circuit in Fig. 1.8 is driven by a sinusoidal source of the form $v_g(t) = \cos\omega_0 t$. Since this is a linear system, its response can be obtained analytically using Laplace transforms. After some algebra, the expressions for the capacitor voltage and the inductor current are found to be

$$x_1(t) = v_C(t) = \frac{\omega_C^2}{\omega_C^2 - \omega_0^2}\left[\cos\omega_0 t - \cos\omega_C t\right] \tag{1.98}$$

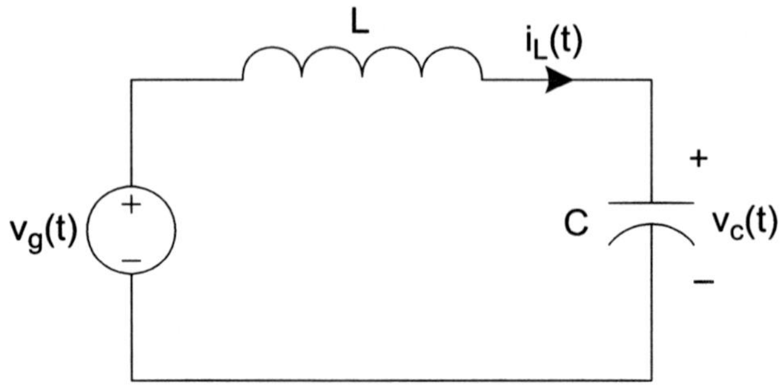

Fig. 1.8. An LC circuit.

and

$$x_2(t) = i_L(t) = \frac{1}{L(\omega_C^2 - \omega_0^2)}\left[\omega_C \sin \omega_C t - \omega_0 \sin \omega_0 t\right] \tag{1.99}$$

where

$$\omega_C = \frac{1}{\sqrt{LC}} \tag{1.100}$$

From (1.98) and (1.99) it is clear that the dynamics of this system involve two different frequencies, ω_0 and ω_C. Depending on how ω_0, L and C are chosen, the following scenarios can arise.

CASE 1. For $L = 1$, $C = 1$ and $\omega_0 = 1.5$ we have

$$\frac{\omega_0}{\omega_C} = \frac{3}{2} \tag{1.101}$$

which means that the frequencies are *commensurate*. As a result, $x_1(t)$ and $x_2(t)$ have the periodic form shown in Figs. 1.9 and 1.10. The attractor for this system is the closed curve shown in Fig. 1.11.

CASE 2. For $L = 2$, $C = 1$ and $\omega_0 = 1.5$ the ratio

$$\frac{\omega_0}{\omega_C} = \frac{3\sqrt{2}}{2} \tag{1.102}$$

is *not* a rational number. In this case, states $x_1(t)$ and $x_2(t)$ exhibit quasi-periodic behavior, as indicated in Figs. 1.12 and 1.13. The attractor is now the rather complicated set shown in Fig. 1.14.

Example 1.9. Let us now consider a slightly more general scenario, where $f(t)$ is a *product* of two sinusoids

$$f(t) = \cos \omega_1 t \cdot \cos \omega_2 t \tag{1.103}$$

As in Example 1.8, $f(t)$ will be periodic when condition (1.96) holds. It is also easily verified that $f(t)$ is quasi-periodic when ω_1 and ω_2 are incommensurate, since

$$f(t) = \frac{1}{2}\cos[(\omega_1 - \omega_2)t] + \frac{1}{2}\cos[(\omega_1 + \omega_2)t] \tag{1.104}$$

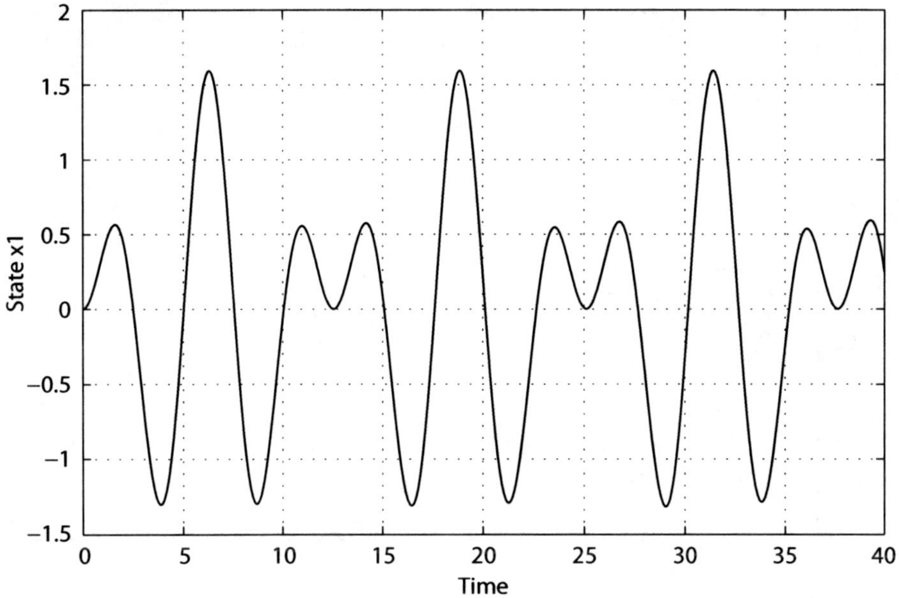

Fig. 1.9. Evolution of state x_1.

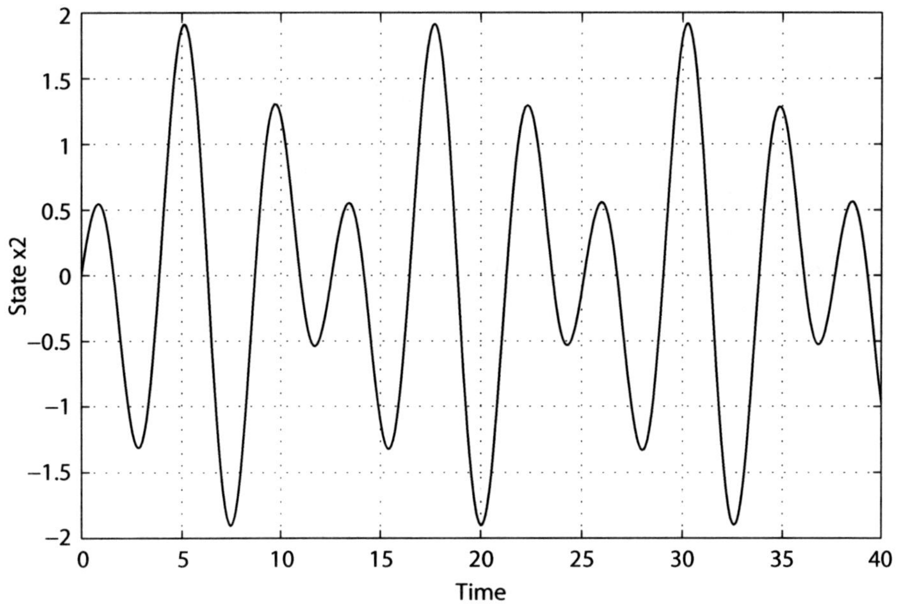

Fig. 1.10. Evolution of state x_2.

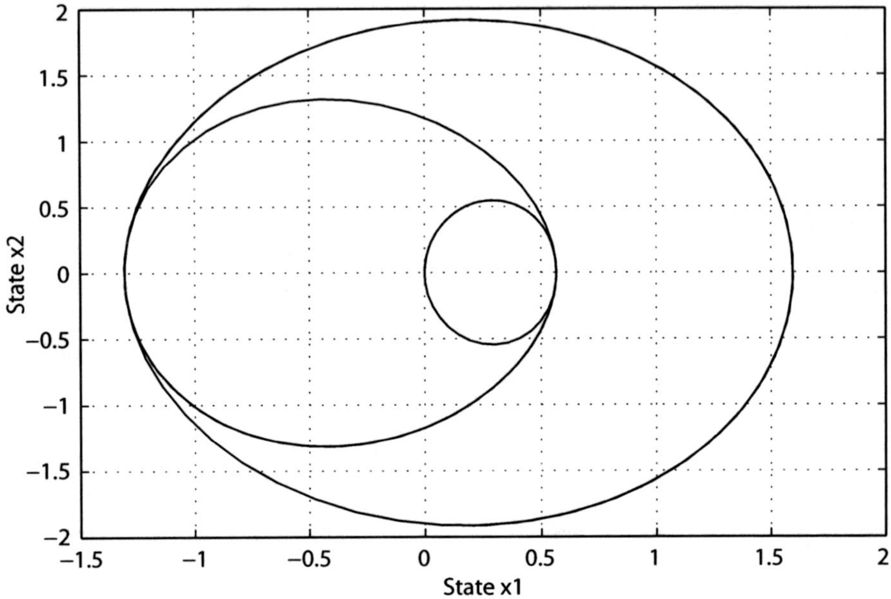

Fig. 1.11. The attractor.

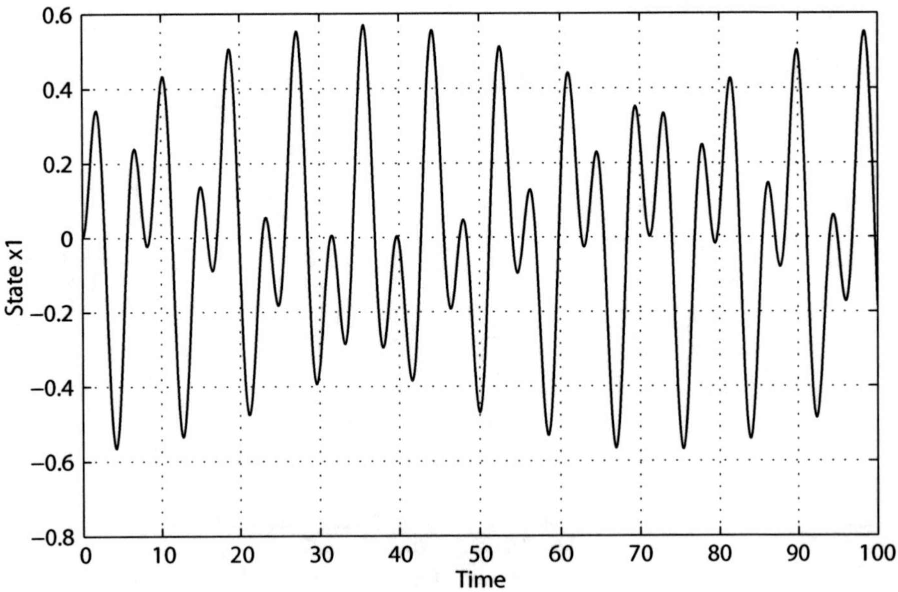

Fig. 1.12. Evolution of state x_1.

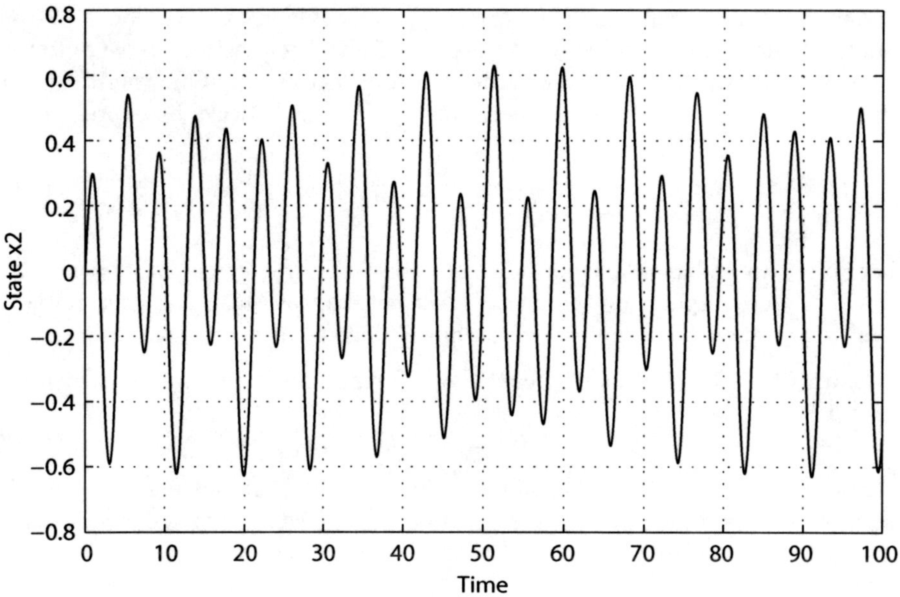

Fig. 1.13. Evolution of state x_2.

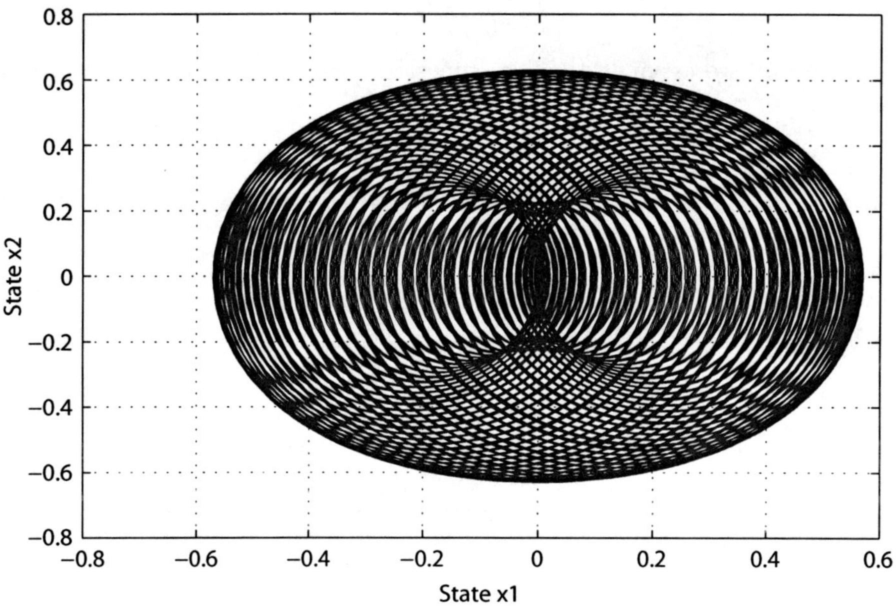

Fig. 1.14. The attractor.

Before proceeding to more general cases, we need to observe that the solutions of nonlinear differential equations are rarely available in analytic form, and are usually obtained numerically. This raises the following question: If we only have a plot of $f(t)$, how can we tell whether it is quasi-periodic? One way to do this is based on the fact that a quasi-periodic function can always be expressed as

$$f(t) = \sum_{i=1}^{N} h_i(t) \tag{1.105}$$

where $h_i(t)$ are independent periodic functions with frequencies ω_i ($i = 1, 2, \ldots, N$). The Fourier spectrum of such a function is always *discrete*, as opposed to an arbitrary aperiodic function which has a continuous spectrum.

Example 1.10. Let us now consider a further generalization, in which $f(t)$ is given as

$$f(t) = a(t) \cdot b(t) \tag{1.106}$$

Functions $a(t)$ and $b(t)$ are assumed to be periodic (but not sinusoidal), with frequencies ω_1 and ω_2, respectively. In this case, we can perform our analysis by expanding $a(t)$ and $b(t)$ in terms of their Fourier series as

$$a(t) = \sum_{n=0}^{\infty} A_n \cos n\omega_1 t \tag{1.107}$$

and

$$b(t) = \sum_{n=0}^{\infty} B_n \cos n\omega_2 t \tag{1.108}$$

(for simplicity we assumed that the functions were even, so only cosine terms appear).

The expansions in (1.107) and (1.108) allow us to express $f(t)$ as

$$f(t) = \sum_{n,k} A_n B_k \cos n\omega_1 t \cdot \cos k\omega_2 t \tag{1.109}$$

and therefore

$$f(t) = \frac{1}{2} \sum_{n,k} A_n B_k \cos[(n\omega_1 + k\omega_2)t] + \frac{1}{2} \sum_{n,k} A_n B_k \cos[(n\omega_1 - k\omega_2)t] \tag{1.110}$$

It is not difficult to show that $f(t)$ will once again be periodic if (1.96) holds, since this implies $n\omega_1 T = 2np\pi$ and $k\omega_2 T = 2kq\pi$ for *all* integers n and k as well. In the incommensurate case, $f(t)$ is guaranteed to be quasiperiodic, since (1.110) is an infinite sum of periodic functions. The frequency spectrum is obviously discrete, with nonzero values at $n\omega_1 \pm k\omega_2$ ($n, k = 0, 1, \ldots$).

An example of a nonlinear system with a quasi-periodic attractor is the forced van der Pol oscillator

$$\begin{aligned} \dot{x}_1 &= x_2 \\ \dot{x}_2 &= (1 - x_1^2)x_2 - x_1 + 0.5 \cos 1.1t \end{aligned} \tag{1.111}$$

This is a *non-autonomous* system, which differs from (1.90) in the sinusoidal forcing term. The time domain responses of $x_1(t)$ and $x_2(t)$ are shown in Figs. 1.15 and 1.16, suggesting that the steady state solutions are *not* periodic.

The phase portrait for this system is provided in Fig. 1.17, which indicates that the quasi-periodic solution is an attractor for this system. It is important

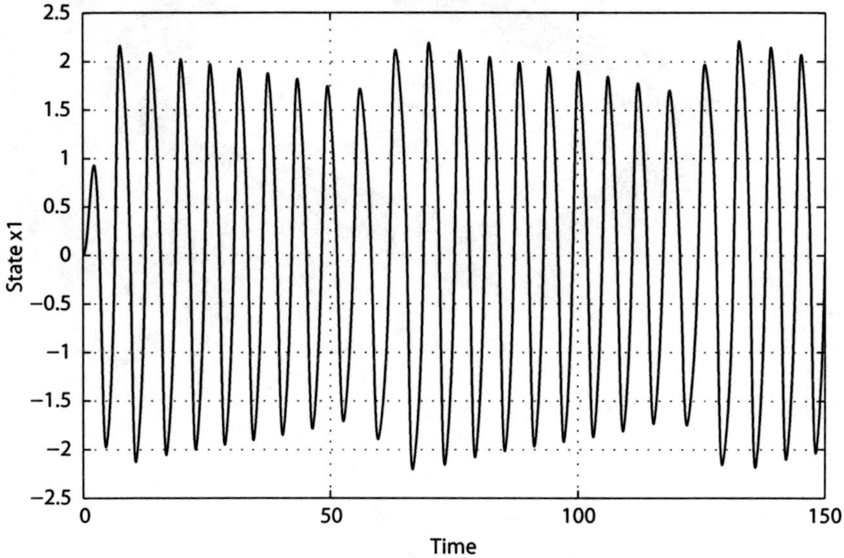

Fig. 1.15. Evolution of state x_1.

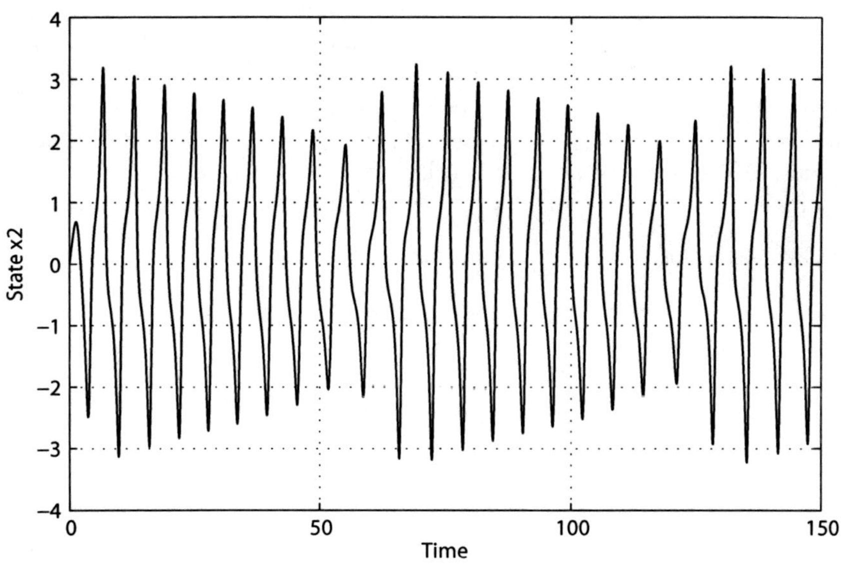

Fig. 1.16. Evolution of state x_2.

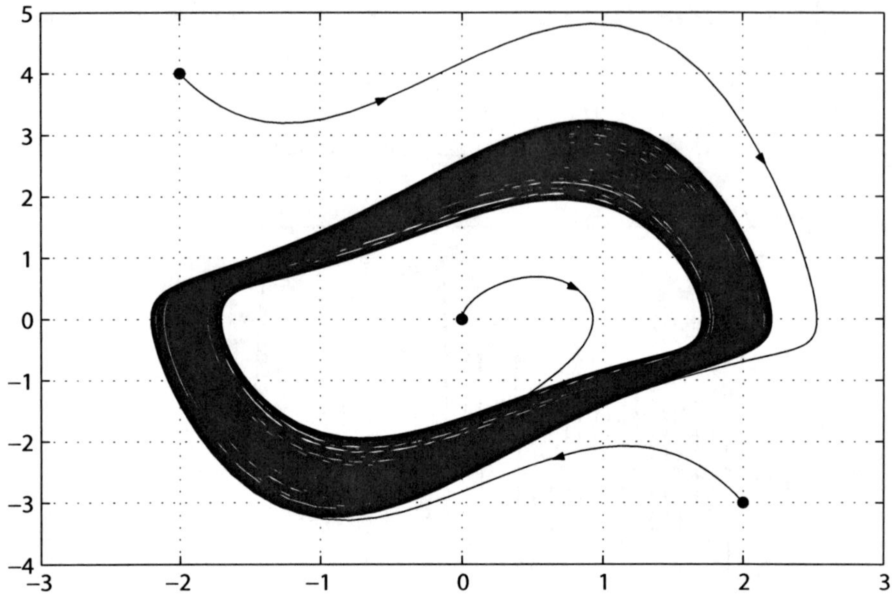

Fig. 1.17. The attractor.

to recognize that a quasi-periodic attractor has a more complicated structure than a limit cycle, which corresponds to a single closed curve in the phase space.

Chaotic (Strange) Attractors

Strange attractors represent the most complicated type of attractor in a nonlinear system. The solutions associated with it appear to be random, and yet they are generated by a well defined deterministic equation. This was a very unexpected (and rather recent) discovery in nonlinear system theory.

We will begin our study of strange attractors with several simple illustrative examples. Most of them are third order systems, since chaos cannot occur in autonomous systems of order < 3 (or non-autonomous systems of order < 2).

Example 1.11. The Lorenz equation

$$\begin{aligned} \dot{x}_1 &= -10x_1 + 10x_2 \\ \dot{x}_2 &= -x_1x_3 + 28x_1 - x_2 \\ \dot{x}_3 &= x_1x_2 - (8/3)x_3 \end{aligned} \tag{1.112}$$

is the first mathematical model in which chaotic behavior was observed. It was introduced in 1963 by Edward Lorenz, as a highly simplified framework for studying atmospheric phenomena.[1] The strange attractor and its projection onto the $x_1 - x_3$ axis are shown in Figs. 1.18 and 1.19, and the evolution of state $x_1(t)$ is provided in Fig. 1.20.

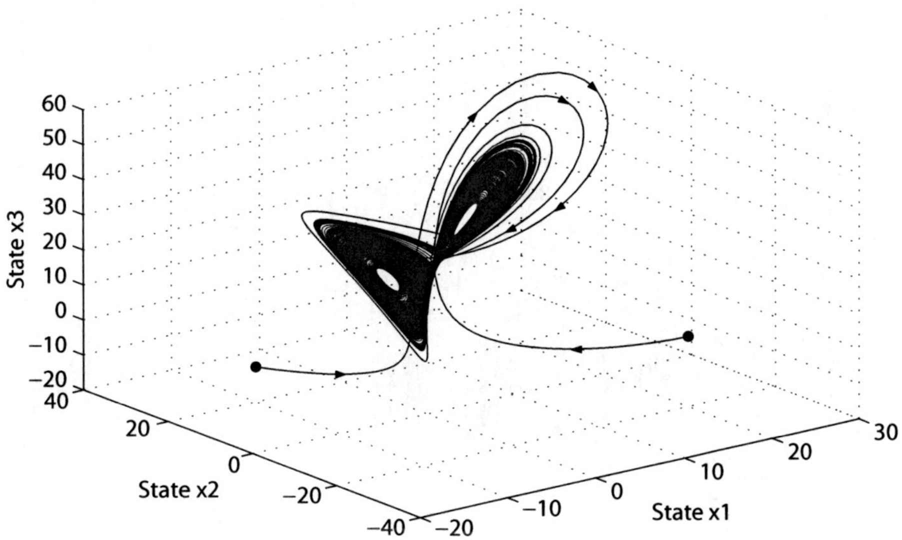

Fig. 1.18. The strange attractor for the Lorenz equation.

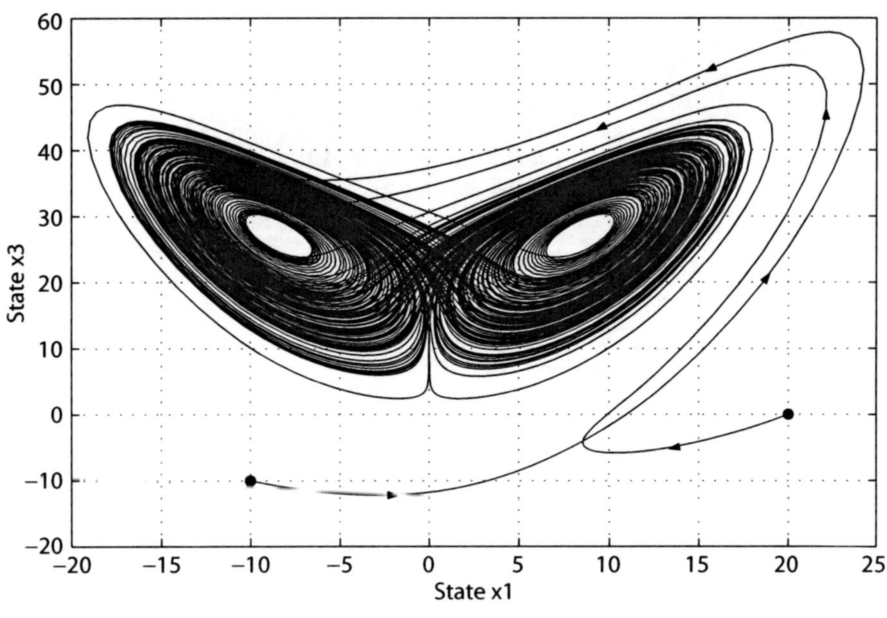

Fig. 1.19. Projection of the attractor onto the $x_1 - x_3$ plane.

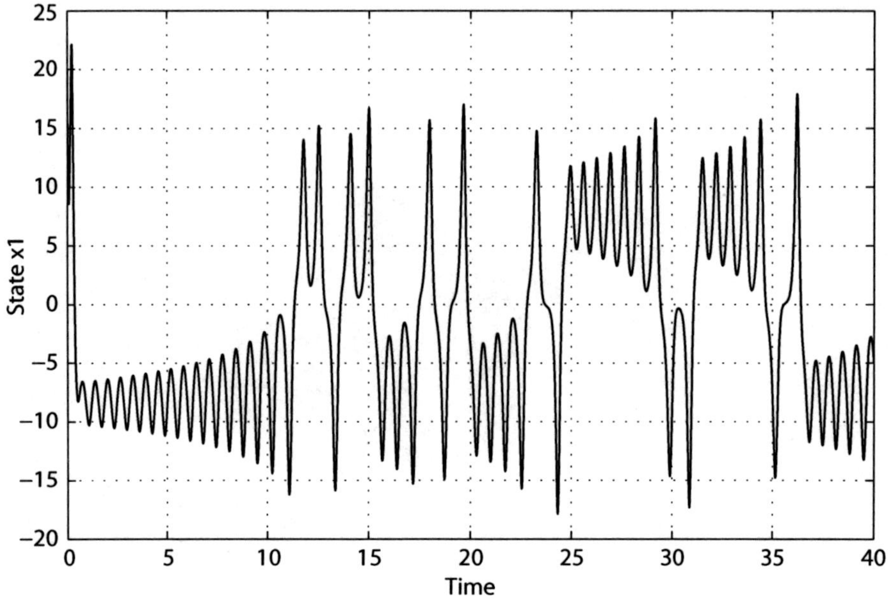

Fig. 1.20. Evolution of state x_1.

Example 1.12. The Rössler equation

$$\begin{aligned}\dot{x}_1 &= -x_2 - x_3 \\ \dot{x}_2 &= x_1 + 0.2x_2 \\ \dot{x}_3 &= 0.2 + x_1 x_3 - 5.7x_3\end{aligned} \tag{1.113}$$

arises in the modeling of fluid flows and chemical reactions, and contains only one nonlinear term.[2] Nevertheless, it has all the characteristics of chaos, as illustrated in Figs. 1.21, 1.22 and 1.23.

Example 1.13. The double-scroll equation

$$\begin{aligned}\dot{x}_1 &= 9x_2 - 9h(x_1) \\ \dot{x}_2 &= x_1 - x_2 + x_3 \\ \dot{x}_3 &= -(100/7)x_2\end{aligned} \tag{1.114}$$

is associated with a class of nonlinear electronic circuits developed by L. O. Chua.[3] The nonlinear term $h(x_1)$ is piecewise-continuous, and is defined as

$$h(x_1) = \begin{cases} (2/7)x_1 - 3/7, & x_1 \geq 1 \\ -(1/7)x_1, & -1 < x_1 < 1 \\ (2/7)x_1 + 3/7, & x_1 \leq -1 \end{cases} \tag{1.115}$$

The attractor and its projection are shown in Figs. 1.24 and 1.25, and the evolution of state $x_1(t)$ is provided in Fig. 1.26.

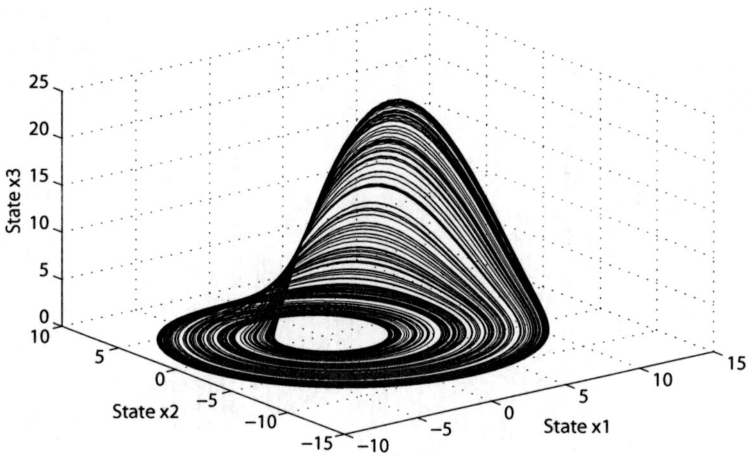

Fig. 1.21. The attractor for the Rössler equation.

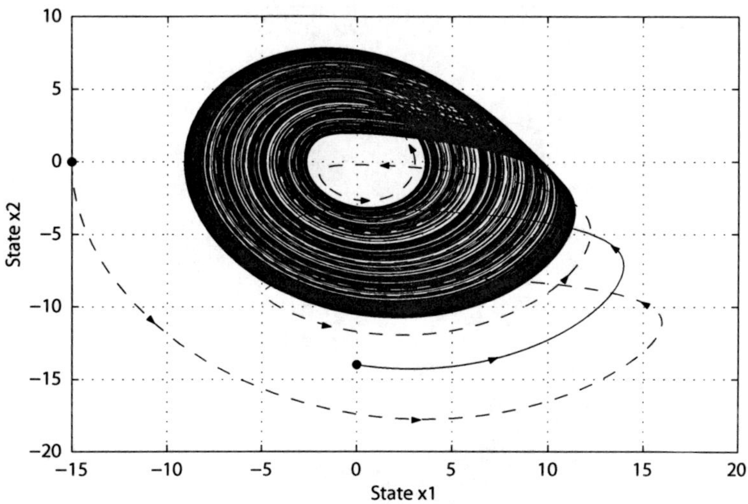

Fig. 1.22. Projection of the attractor onto the $x_1 - x_2$ plane.

Example 1.14. The forced Duffing equation

$$
\begin{aligned}
\dot{x}_1 &= x_2 \\
\dot{x}_2 &= x_1 - x_1^3 - 0.25x_2 + 0.3\cos t
\end{aligned}
\tag{1.116}
$$

provides an example of chaotic behavior in non-autonomous systems.[4] The attractor and state $x_1(t)$ are shown in Figs. 1.27 and 1.28.

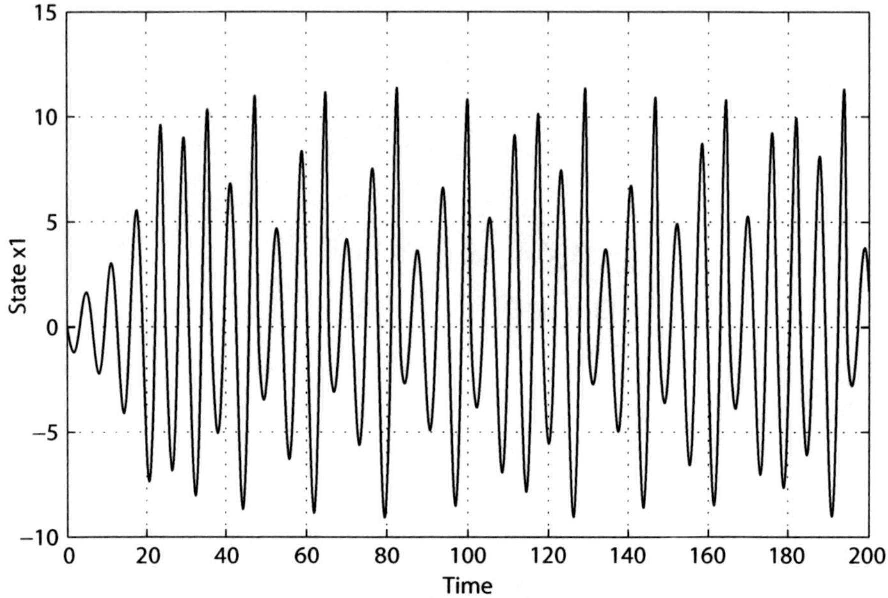

Fig. 1.23. Evolution of state x_1.

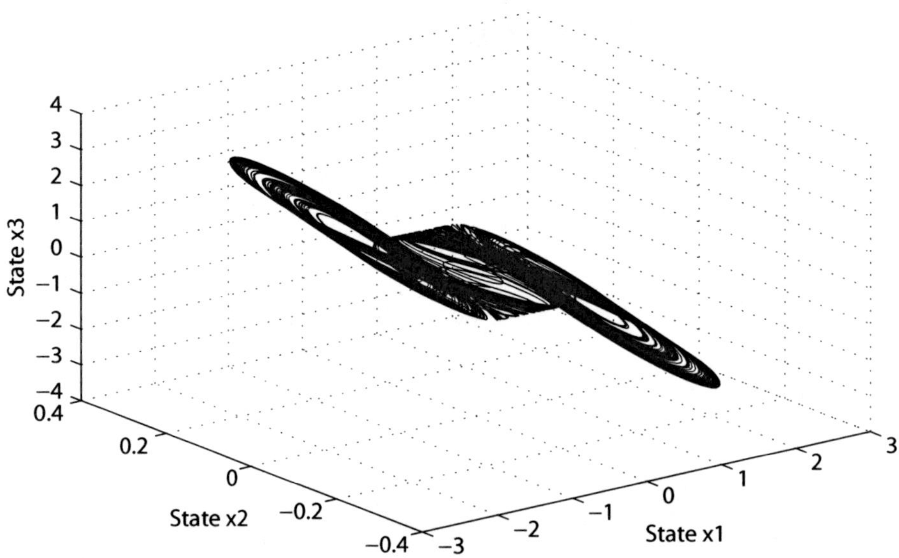

Fig. 1.24. The attractor for the double scroll equation.

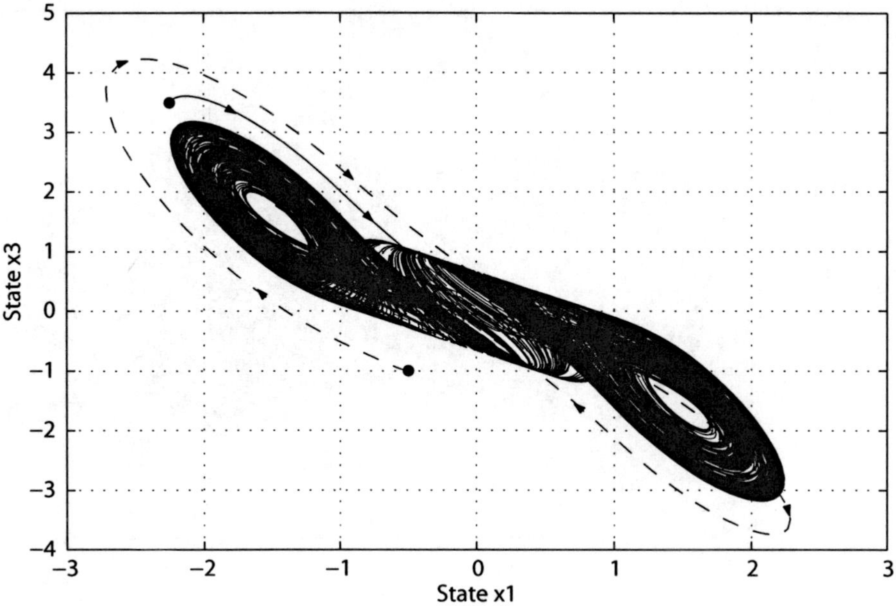

Fig. 1.25. Projection of the attractor onto the $x_1 - x_3$ plane.

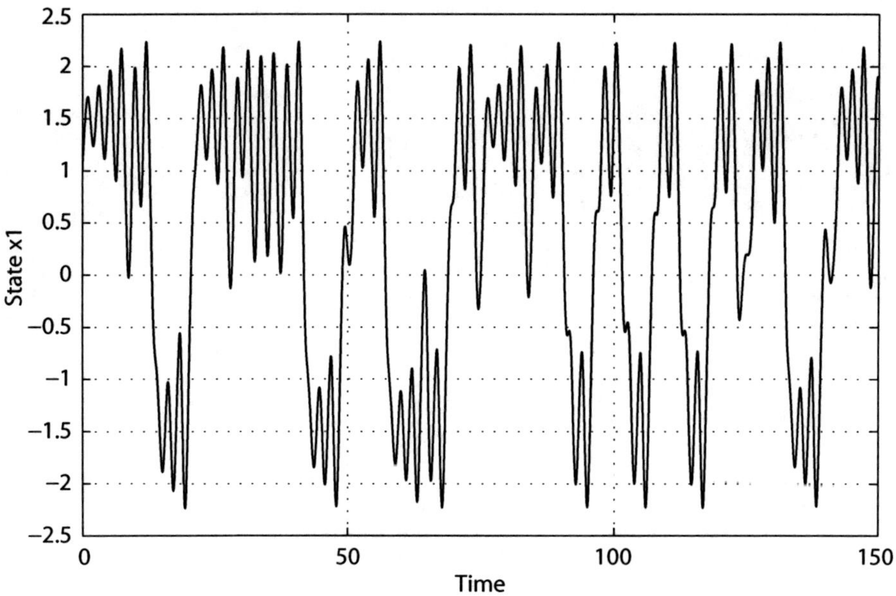

Fig. 1.26. Evolution of state x_1.

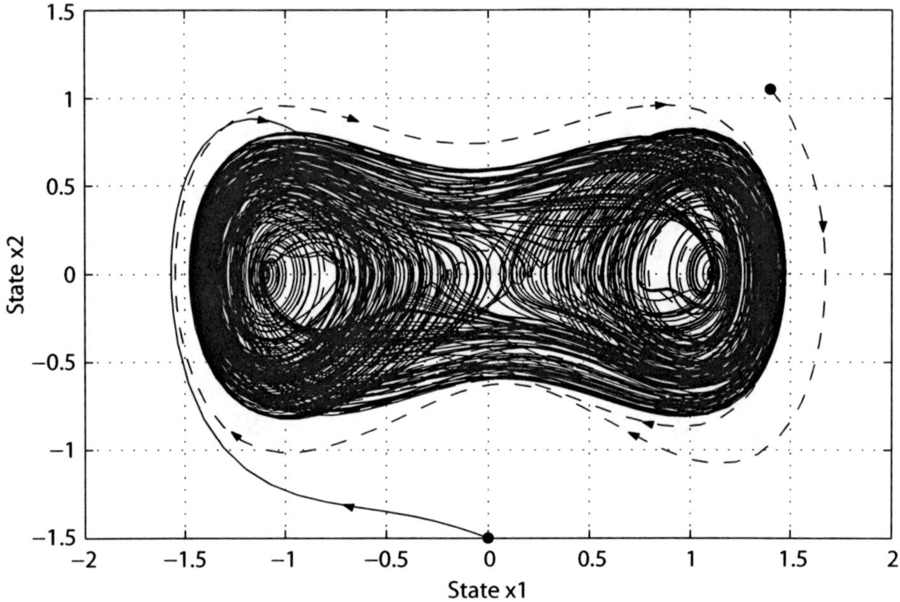

Fig. 1.27. The attractor for the forced Duffing equation.

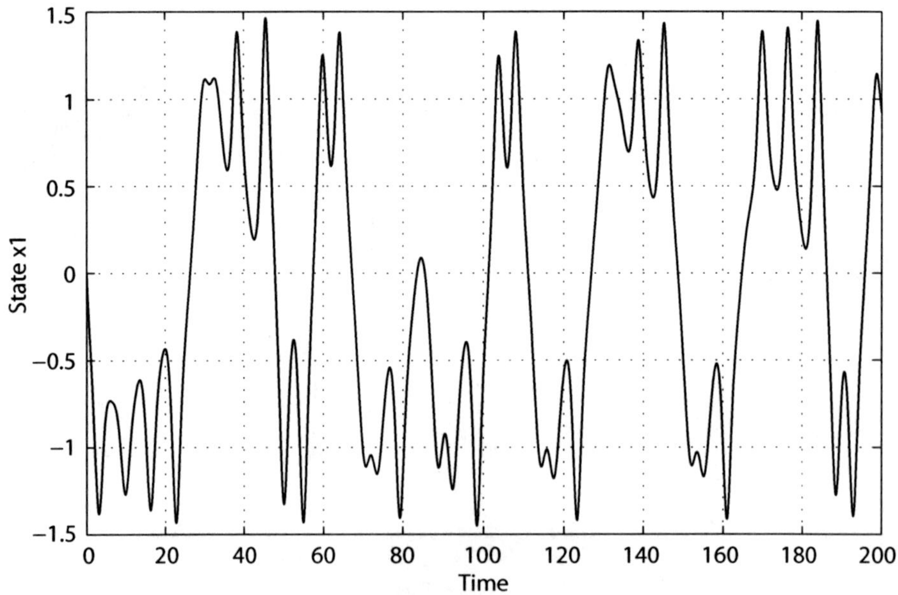

Fig. 1.28. Evolution of state x_1.

1.3 Trademarks of Chaos

In addition to the strange and complicated shape of the attractor, there are a number of other features that are unique to chaotic behavior. In this section, we will focus on the three most prominent ones.

Non-Periodic Behavior

Chaotic trajectories are *not* periodic, but are bounded for all t. Several solutions with this property were shown in Figs. 1.20, 1.23 and 1.26 of the previous section. A natural question that arises in this context is how one can be sure that a trajectory is chaotic and not quasi-periodic (or perhaps periodic with a very long period). A plot of the state variables as a function of time is not particularly reliable in that respect. Instead, it is better to observe the frequency spectrum of the trajectory. In the periodic and quasi-periodic cases the spectrum is necessarily *discrete*, while the chaotic one is *continuous*.

Sensitivity to Initial Conditions

Perhaps the most important characteristic of chaotic trajectories is their sensitivity to initial conditions. For all 'normal' attractors, trajectories that start from similar initial conditions remain close together at all times. The following three examples clearly illustrate this point.

Example 1.15. The predator-prey model (1.79) is an example of a system in which the attractors are *equilibria*. In Fig. 1.29, we show two trajectories whose initial conditions differ by 2%. It is clear from the graph that the two solutions remain close at all times, and ultimately merge as they approach the equilibrium.

Example 1.16. For the van der Pol oscillator (1.90), the attractor is a *limit cycle*. A pair of trajectories whose initial conditions differ by 20% are shown in Fig. 1.30. It is readily observed that the difference between the two solutions is small, and becomes negligible in the steady state.

Example 1.17. The forced van der Pol oscillator (1.111) has a *quasi-periodic* attractor. In Fig. 1.31 we show two of its solutions with initial conditions that differ by 10%. As in the previous examples, the trajectories remain close at all times, although in this case they never become completely indistinguishable.

In contrast to Examples 1.15 - 1.17, chaotic solutions exhibit exponential sensitivity with respect to initial conditions. To see this, let us consider two solutions of system (1.114), whose initial conditions differ by only 0.01%. The evolution of the two trajectories is shown in Fig. 1.32, indicating that after a sufficiently long time there is no resemblance between them. It is important to recognize, however, that this type of system exhibits only *eventual* unpredictability - the trajectories are actually very similar in the short term.

How can we explain such extreme sensitivity with respect to initial conditions? One possible interpretation is based on so-called *Lyapunov exponents*, which we now briefly describe.

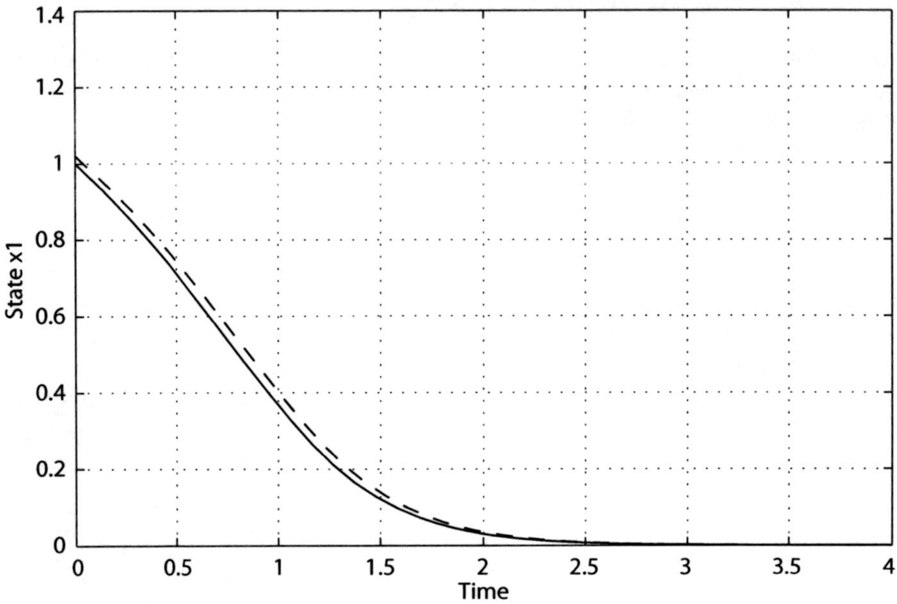

Fig. 1.29: Two initially close trajectories for the predator-prey model (the attractor for this system is an equilibrium).

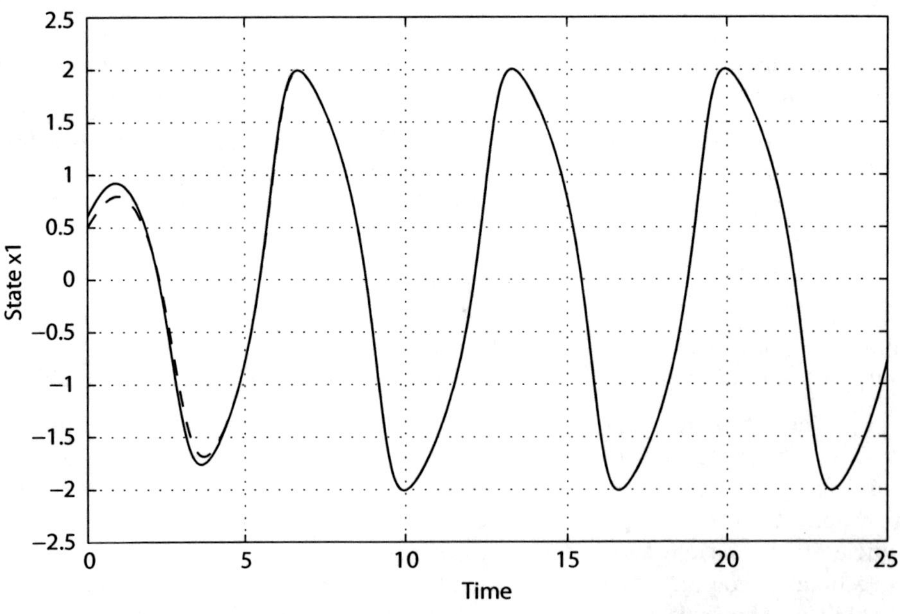

Fig. 1.30: Two initially close trajectories for the van der Pol oscillator (the attractor for this system is a limit cycle).

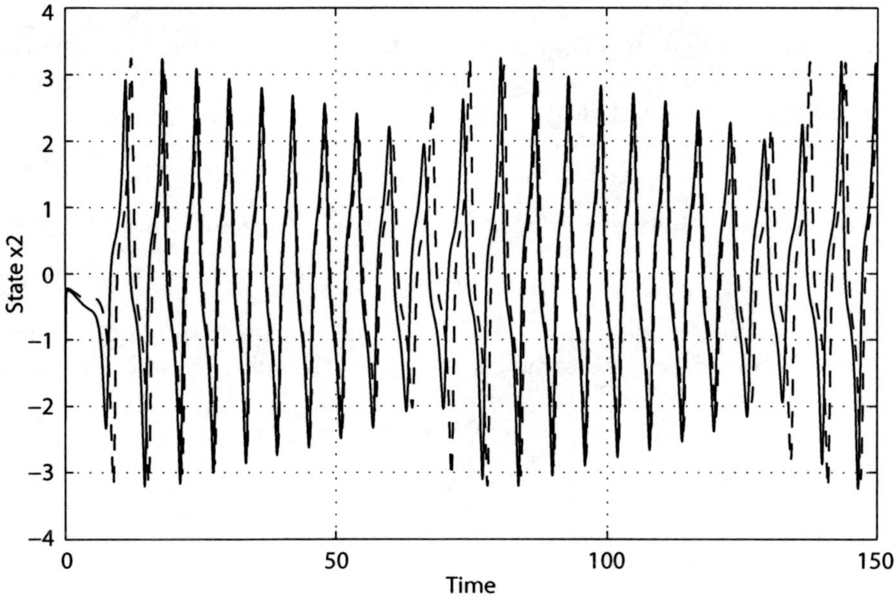

Fig. 1.31: Two initially close trajectories for the forced van der Pol oscillator (this system has a quasi-periodic attractor).

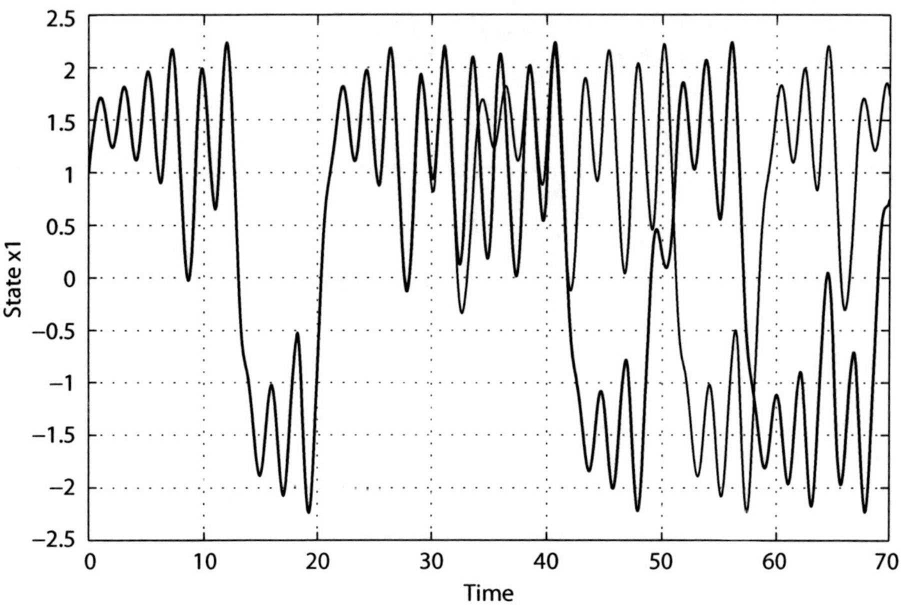

Fig. 1.32: Two initially close trajectories for the double scroll model (this system has a strange attractor).

Lyapunov Exponents

Suppose that we have two trajectories $x(t; t_0, x_0)$ and $\tilde{x}(t; t_0, x_0 + \delta x_0)$, whose initial conditions differ by some small δx_0. In the following, we will be interested in estimating how the difference

$$\delta x(t) = \tilde{x}(t; t_0, x_0 + \delta x_0) - x(t; t_0, x_0) \tag{1.117}$$

evolves over time. Since δx_0 is small, we can assume that $\delta x(t)$ has the general form

$$\delta x(t) = \Phi(t; t_0, x_0)\delta x_0 \tag{1.118}$$

where $\Phi(t; t_0, x_0)$ is a time-varying matrix that needs to be determined. In order to compute this matrix, we should first observe that differentiation of (1.118) yields

$$\delta \dot{x}(t) = \dot{\Phi}(t; t_0, x_0)\delta x_0 \tag{1.119}$$

Furthermore, since

$$\dot{x}(t) = f(x, t) \tag{1.120}$$

and

$$\dot{\tilde{x}}(t) = f(\tilde{x}, t) \tag{1.121}$$

we also know that

$$\delta \dot{x}(t) = f(\tilde{x}, t) - f(x, t) \approx J(x(t), t)\delta x(t) \tag{1.122}$$

where $J(x(t), t)$ is the Jacobian of $f(x, t)$ evaluated at $x(t; t_0, x_0)$. This linearization is appropriate for as long as $\delta x(t)$ remains small.[5]

Replacing $\delta x(t)$ with the expression (1.118), we obtain

$$\delta \dot{x}(t) = J(x(t), t)\Phi(t; t_0, x_0)\delta x_0 \tag{1.123}$$

Equating (1.119) and (1.123) now produces the so-called *variational equation*

$$\dot{\Phi}(t; t_0, x_0) = J(x(t), t)\Phi(t; t_0, x_0) \tag{1.124}$$

Note that the initial condition for this equation is

$$\Phi(t_0; t_0, x_0) = I \tag{1.125}$$

since $\delta x(t_0) = \delta x_0$.

Once $x(t; t_0, x_0)$ is known, we can compute $J(x(t), t)$ and solve (1.124) numerically to obtain $\Phi(t; t_0, x_0)$. The eigenvalues of this matrix (denoted $m_i(t)$) are functions of time, and are used to define Lyapunov exponents as

$$\lambda_i = \lim_{t \to \infty} \frac{1}{t} \ln |m_i(t)| \tag{1.126}$$

Remark 1.3. In cases when $J(x(t), t)$ has eigenvalues in the right half-plane (which is typical for chaotic behavior), the solution of (1.124) can pose serious numerical problems, since $\Phi(t; t_0, x_0) \to \infty$ when $t \to \infty$. As a result, Lyapunov exponents are usually not computed by direct integration.

ATTRACTOR	LYAPUNOV EXPONENTS
Equilibrium point	$0 > \lambda_1 \geq \ldots \geq \lambda_n$
Limit Cycle	$\lambda_1 = 0$ $0 \geq \lambda_2 \geq \ldots \geq \lambda_n$
Quasi-Periodic (2 frequencies)	$\lambda_1 = \lambda_2 = 0$ $0 \geq \lambda_3 \geq \ldots \geq \lambda_n$
Quasi-Periodic (k frequencies)	$\lambda_1 = \ldots = \lambda_k = 0$ $0 \geq \lambda_{k+1} \geq \ldots \geq \lambda_n$
Chaotic	$\lambda_1 > 0, \quad \sum_{i=1}^{n} \lambda_i < 0$

Table 1.3. A characterization of attractors based on Lyapunov exponents.

What is the benefit of Lyapunov exponents? For one thing, they allow us to 'neatly' classify attractors in the manner shown in Table 1.3. Lyapunov exponents also provide an intuitive explanation for the sensitivity of chaotic trajectories with respect to initial conditions. Namely, since $\lambda_1 > 0$, equation (1.118) implies that any two nearby trajectories will diverge at an average rate of $e^{\lambda_1 t}$. It should be noted, however, that the distance between the trajectories is ultimately bounded by the dimensions of the attractor (which are finite). Since strange attractors typically exhibit spatial 'folding' (see Fig. 1.21 for a typical illustration), initially divergent trajectories can in fact get relatively close together after a certain amount of time, only to drift apart again.

Fractal Dimension

Another unique characteristic of chaotic attractors is related to their dimensionality. Our intuitive understanding of this property is usually Euclidean. We tend to think of a line as 1-dimensional, a surface as 2-dimensional, a volume as 3-dimensional, etc. In this context, dimensionality serves as a measure of the complexity of a geometric object (for example, we can legitimately claim that a surface is more 'complex' than a line).

Since strange attractors are generally very complicated geometrical structures, it would clearly be interesting to establish their dimension. This would enable us to categorize a particular strange attractor as a line, surface, volume, or perhaps none of the above... In order to do this, we first need a definition of dimension which is broader than the traditional Euclidean one, and can be applied to arbitrary sets of points. While there are a number of such definitions, in the following we will focus on one which is known as *capacitive dimension*.

To understand this concept, imagine that the entire state space is divided into 'cubes' with sides of length ε. If $N(\varepsilon)$ denotes the number of cubes that contain points from the set that we are examining, we can introduce the quantity $C(\varepsilon)$, defined as

$$C(\varepsilon) = \frac{\ln N(\varepsilon)}{\ln(1/\varepsilon)} \tag{1.127}$$

The limit

$$d_C = \lim_{\varepsilon \to 0} C(\varepsilon) \tag{1.128}$$

is known as the capacitive dimension of this set. The following three examples show that the definition of capacitive dimension is consistent with the Euclidean one.

Example 1.18. Let us begin by considering a single point, since this is the simplest possible geometrical object. As shown in Fig. 1.33, a single box is sufficient to cover the point for any choice of ε. This implies that $N(\varepsilon) = 1$ and therefore $C(\varepsilon) = 0$ as well, which is consistent with the Euclidean dimension.

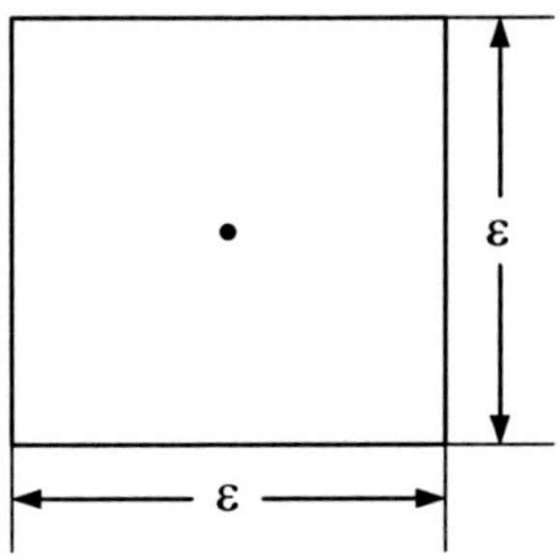

Fig. 1.33. The dimensionality of a point.

Example 1.19. Consider the straight line of length L shown in Fig. 1.34.

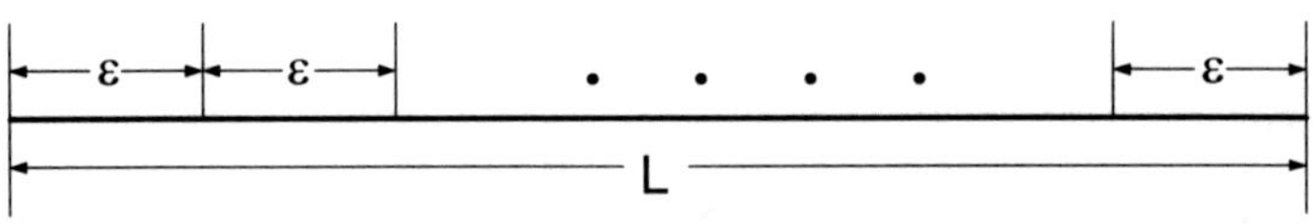

Fig. 1.34. The dimensionality of a line.

The number of 'boxes' needed to cover this set of points is

$$N(\varepsilon) = \frac{L}{\varepsilon} \tag{1.129}$$

and therefore

$$C(\varepsilon) = \frac{\ln N(\varepsilon)}{\ln(1/\varepsilon)} = \frac{\ln L}{\ln(1/\varepsilon)} + \frac{\ln(1/\varepsilon)}{\ln(1/\varepsilon)} = 1 + \frac{\ln L}{\ln(1/\varepsilon)} \tag{1.130}$$

Since

$$\ln(1/\varepsilon) \to \infty \text{ when } \varepsilon \to 0 \tag{1.131}$$

it follows that

$$d_C = \lim_{\varepsilon \to 0} C(\varepsilon) = 1 \tag{1.132}$$

which obviously matches the Euclidean dimension.

Example 1.20. Consider a square surface with sides of length L.

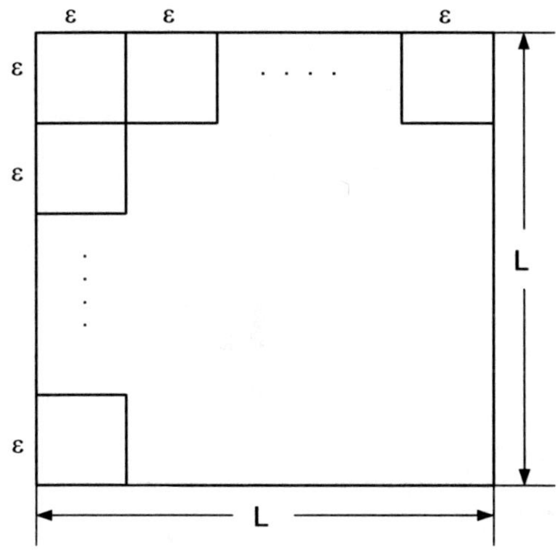

Fig. 1.35. The dimensionality of a square.

The number of boxes needed to cover this surface is

$$N(\varepsilon) = \frac{L^2}{\varepsilon^2} \tag{1.133}$$

which means that

$$C(\varepsilon) = \frac{\ln L^2}{\ln(1/\varepsilon)} + \frac{\ln(1/\varepsilon)^2}{\ln(1/\varepsilon)} = 2 + \frac{\ln L^2}{\ln(1/\varepsilon)} \tag{1.134}$$

and therefore

$$d_C = \lim_{\varepsilon \to 0} C(\varepsilon) = 2 \tag{1.135}$$

Once again we observe consistency with the Euclidean definition.

To see how the definition of capacitive dimension extends to some less conventional sets, let us consider a line of unit length, which we will transform through an infinite sequence of steps. The transformation consists of removing the middle third of each existing segment, until we arrive at a disconnected set of points (when $n \to \infty$). This procedure is illustrated in Fig. 1.36, and the resulting set is known as the *Cantor set*.

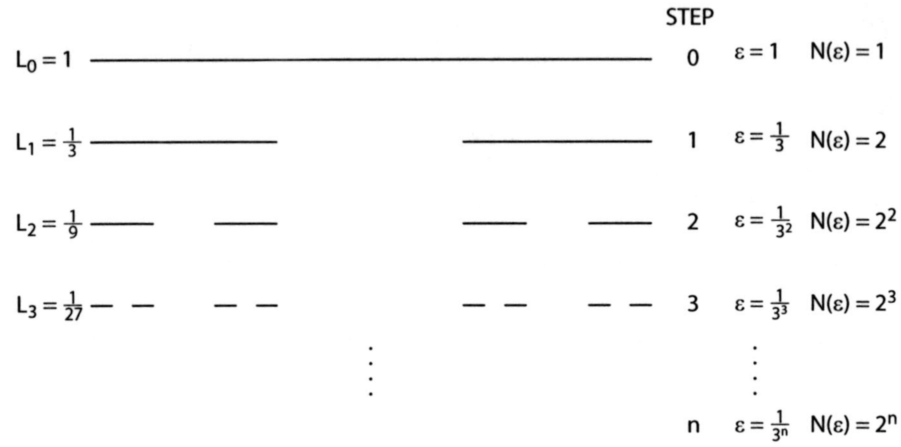

Fig. 1.36. The construction of the Cantor set.

To figure out the capacitive dimension of the Cantor set, let us observe that we need a single box of size $\varepsilon = 1$ to cover the set in Step 0, 2 boxes of size $\varepsilon = 1/3$ to do this in Step 1, 4 boxes of size $\varepsilon = 1/9$ in Step 2, etc. Obviously, in Step n, we will need 2^n boxes of size $\varepsilon = (1/3)^n$. Since $\varepsilon \to 0$ when $n \to \infty$, we have

$$d_C = \lim_{\varepsilon \to 0} \left[\frac{\ln N(\varepsilon)}{\ln(1/\varepsilon)} \right] = \lim_{n \to \infty} \left[\frac{\ln 2^n}{\ln 3^n} \right] = 0.63 \tag{1.136}$$

This is a somewhat surprising result, which suggests that there are certain sets with non-integer dimensionality. Sets of this type are known as *fractals*.

How does this apply to strange attractors? Well, it turns out that they too are fractals, unlike all other types of attractors (which have integer dimension). In view of that, what would it mean if we established that a strange attractor has $d_C = 2.4$? This number would indicate that the attractor is more complex than a surface, but less complex than a volume. Apart from dimensionality, fractals have a number of other interesting properties. Perhaps the most striking one is the fact that they possess both a macroscopic and microscopic structure. The 'pattern' exhibited by the attractor in the state space is replicated on *all* scales of magnitude.

1.4 Bifurcations

Chaos is not a universal phenomenon, and occurs only under certain conditions. To understand what this means, let us consider a variant of the double-scroll equation

$$\begin{aligned} \dot{x}_1 &= a[x_2 - h(x_1)] \\ \dot{x}_2 &= x_1 - x_2 + x_3 \\ \dot{x}_3 &= -bx_2 \end{aligned} \tag{1.137}$$

where a and b are parameters than can change (parameter variations are very common in engineering models). In the following, we will look at three different scenarios.

CASE 1. For $a = 8.8$ and $b = 16$ the system has a periodic solution with a *single* amplitude. This property is illustrated in Fig. 1.37, where we show the evolution of state $x_1(t)$.

CASE 2. When $a = 9.1$ and $b = 16$, the solution is still periodic, but with *two* different amplitudes. This type of change is known as *period doubling*, and is shown in Fig. 1.38.

CASE 3. If parameter a is further increased to $a = 9.3$ (with b still fixed at 16), the system exhibits chaotic behavior. The corresponding trajectory for $x_1(t)$ is shown in Fig. 1.39, and the attractor is provided in Fig. 1.40.

Our analysis of the double-scroll equation suggests that the emergence of chaos should be studied in the context of system parameters. This basically means that a dynamic system should be described as

$$\dot{x} = f(x, p, t) \tag{1.138}$$

where p is a vector of parameters (for the model in (1.137), this vector would be $p = [a \quad b]^T$). Given such a description, our objective will be to analyze the system dynamics as vector p varies. In this analysis, values of p for which the system behavior changes qualitatively will prove to be particularly important, and we will refer to them as *bifurcation points*. The following examples illustrate different types of bifurcations that can occur in simple nonlinear systems.

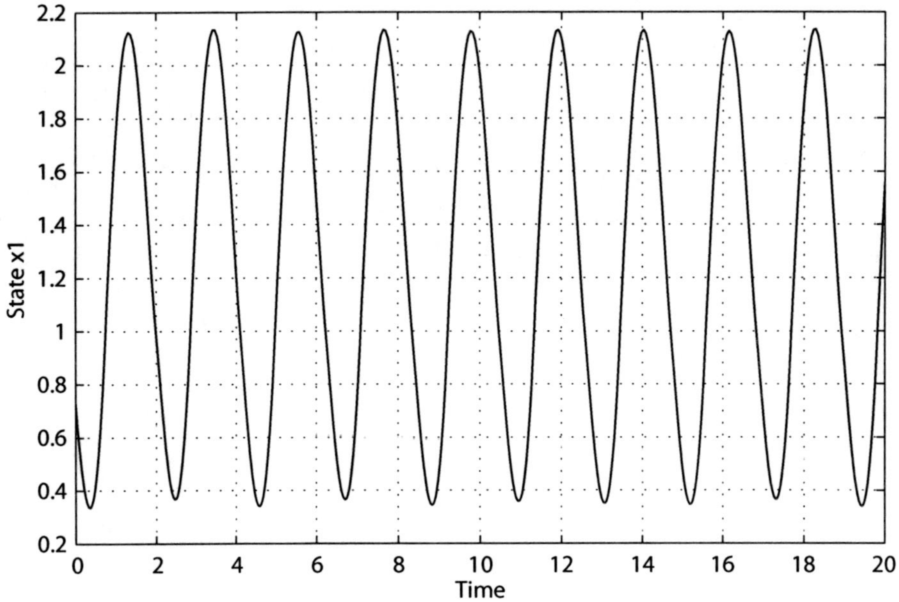

Fig. 1.37. Periodic solution with a single amplitude.

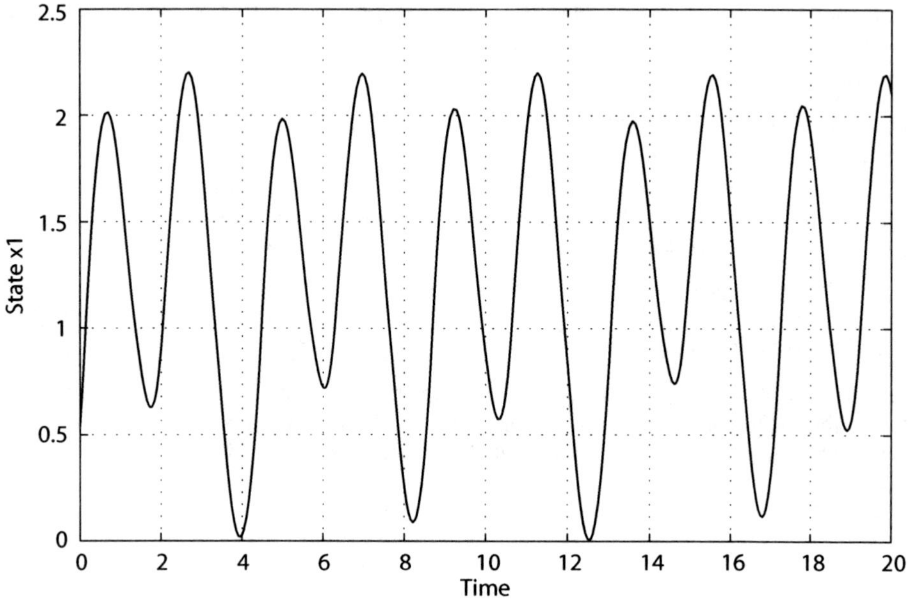

Fig. 1.38. Periodic solution with two amplitudes.

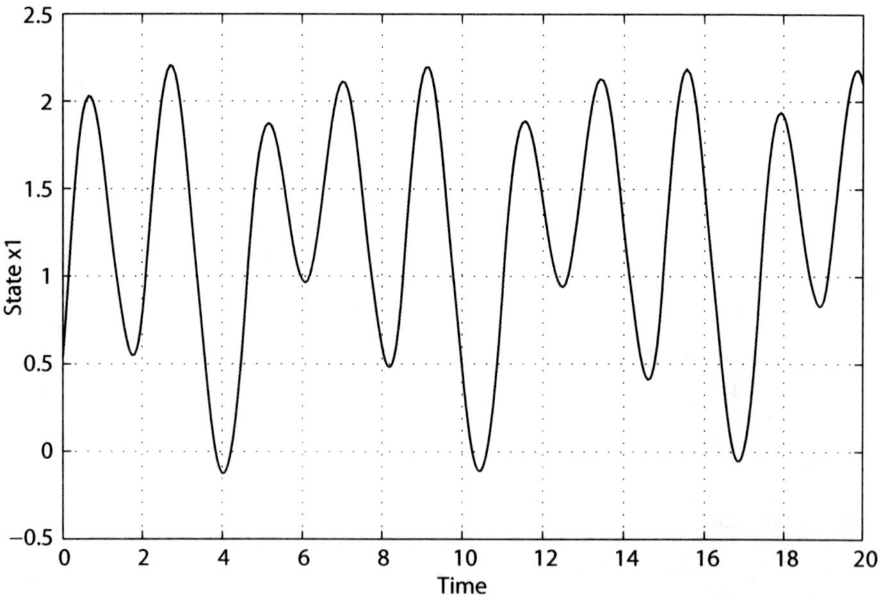

Fig. 1.39. A chaotic solution.

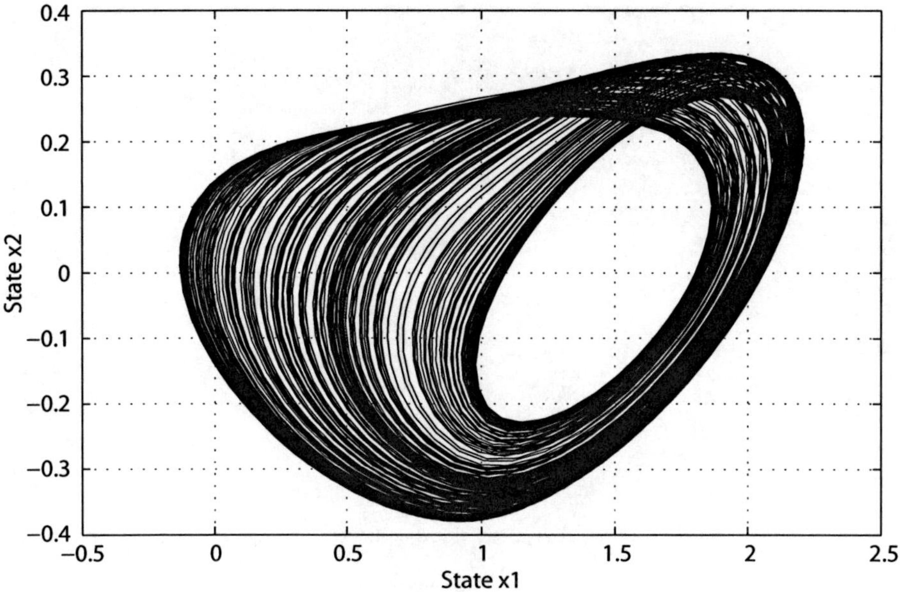

Fig. 1.40. The strange attractor.

Example 1.21. (Saddle-node bifurcation). Consider the system

$$\dot{x} = p - x^2 \tag{1.139}$$

where p is a scalar parameter. For values $p < 0$, this system has no equilibrium, since equation

$$p - x^2 = 0 \tag{1.140}$$

has no real solutions. For $p = 0$, there is a unique equilibrium at $x = 0$, and for $p > 0$ there are two equilibria corresponding to $\pm\sqrt{p}$. Note that the linearized system at $x^e = \sqrt{p}$ is

$$\dot{y} = -2\sqrt{p}\,y \tag{1.141}$$

which indicates that this is a *stable* equilibrium. On the other hand, the linearization at $x^e = -\sqrt{p}$ produces

$$\dot{y} = 2\sqrt{p}\,y \tag{1.142}$$

which is clearly *unstable*.

The analysis outlined above is best summarized by plotting the system equilibria as a function of p. Such a plot is known as a *bifurcation diagram*, and is very useful in analyzing chaos. The bifurcation diagram for the system (1.139) is shown in Fig. 1.41, in which unstable equilibria are indicated by dashed lines.

From Fig. 1.41 it is quite obvious that this system undergoes a qualitative change for $p = 0$, when two new equilibria emerge. As a result, $p = 0$ represents a bifurcation point for this system. Since one of the new equilibria is always stable and the other one unstable, this type of bifurcation is known as a *saddle-node*.

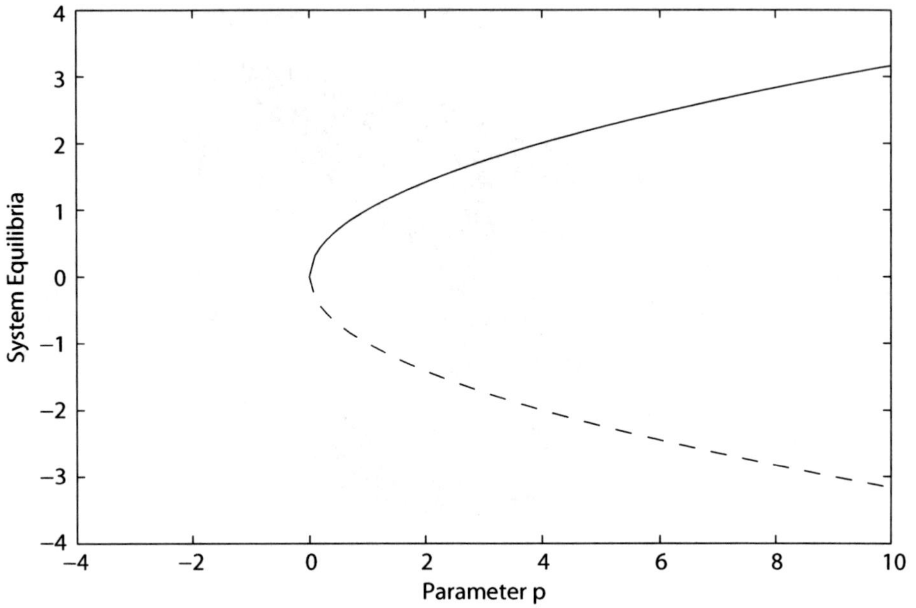

Fig. 1.41. A saddle node bifurcation.

Example 1.22. (Pitchfork bifurcation). Consider the system

$$\dot{x} = x(p - x^2) \tag{1.143}$$

For $p < 0$, the only equilibrium is $x = 0$, while for $p > 0$ there are two more, corresponding to $\pm\sqrt{p}$. The Jacobian associated with (1.143) has the form

$$J(x) = p - 3x^2 \tag{1.144}$$

For the three equilibria, we have the following scenarios:

(*i*) For $x^e = 0$, the linearized system is

$$\dot{x} = px \tag{1.145}$$

Consequently, this equilibrium is stable when $p < 0$, and unstable when $p > 0$.

(*ii*) For $x^e = \sqrt{p}$, the linearized system is

$$\dot{y} = -2py \tag{1.146}$$

where $y = x - x^e$. Since this equilibrium exists only for $p > 0$, it is always stable.

(*iii*) For $x^e = -\sqrt{p}$, the linearized system is exactly the same as (1.146), and the same analysis applies. As a result, this equilibrium is stable for all $p > 0$.

The above analysis is summarized in the bifurcation diagram shown in Fig. 1.42.

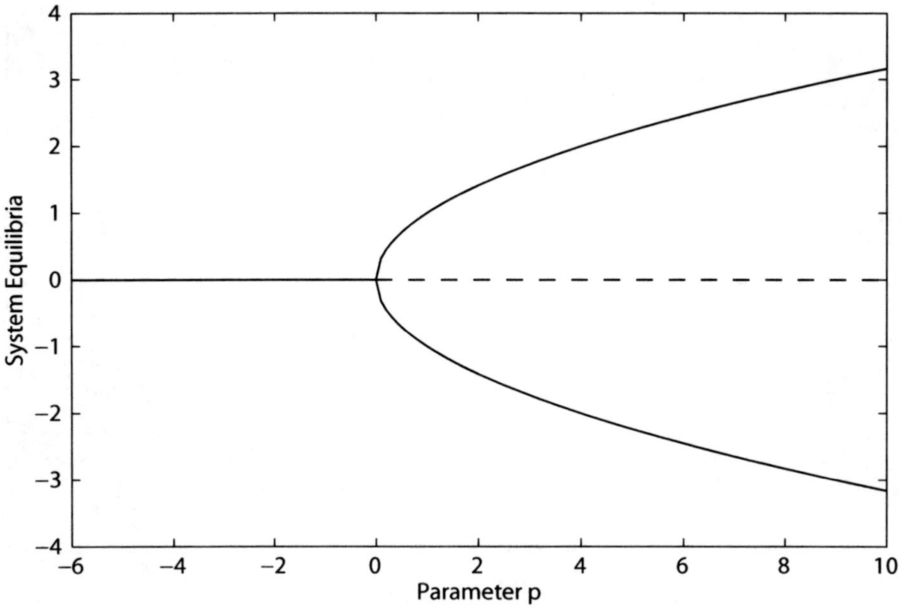

Fig. 1.42. A pitchfork bifurcation.

It is clear that $p = 0$ represents a bifurcation point for this system. This type of bifurcation where two *stable* equilibria emerge is known as a *pitchfork* (the reason for this name is obvious from Fig. 1.42).

Example 1.23. (Transcritical bifurcation). The system

$$\dot{x} = x(p - x) \tag{1.147}$$

has two equilibria for all values of p, located at $x^e = 0$ and $x^e = p$, respectively. Linearization around these equilibria gives rise to the following two cases:

(i) For $x^e = 0$, the linearized system is

$$\dot{x} = px \tag{1.148}$$

Consequently, this equilibrium is stable when $p < 0$, and unstable when $p > 0$.

(ii) For $x^e = p$, the linearized system is

$$\dot{y} = -py \tag{1.149}$$

where $y = x - x^e$. This equilibrium is stable when $p > 0$, and unstable when $p < 0$.

The bifurcation diagram for this system is shown in Fig. 1.43. It indicates that $p = 0$ is once again a bifurcation point, but the nature of the qualitative change is different from the previous two examples. In this case no new equilibria are created following the bifurcation, but there is an exchange of stability between the two equilibria. Such a bifurcation is referred to as *transcritical*.

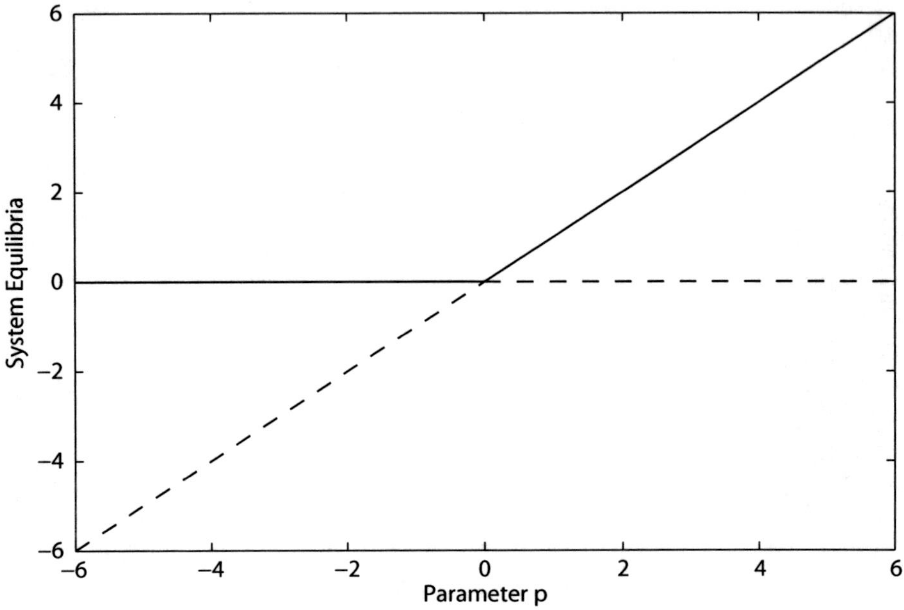

Fig. 1.43. A transcritical bifurcation.

Bifurcations are by no means limited to equilibria. The following example describes the so-called Hopf bifurcation, in which a limit cycle is born for a particular parameter value.

Example 1.24. (Hopf bifurcation). Consider the system

$$\begin{aligned} \dot{x}_1 &= p_1 x_1 - p_2 x_2 - (x_1^2 + x_2^2) x_1 \\ \dot{x}_2 &= p_2 x_1 + p_1 x_2 - (x_1^2 + x_2^2) x_2 \end{aligned} \tag{1.150}$$

which contains two parameters, p_1 and p_2. It is easily verified that this system has an equilibrium at the origin, and the corresponding Jacobian is

$$J(x^e) = \begin{bmatrix} p_1 & -p_2 \\ p_2 & p_1 \end{bmatrix} \tag{1.151}$$

The eigenvalues of $J(x^e)$ are $\lambda_{1,2} = p_1 \pm jp_2$, which means that this equilibrium is stable for $p_1 < 0$ and unstable for $p_1 > 0$.

It is important to recognize that the stability of the equilibrium is not the only thing that changes when p_1 passes through zero. To see this more clearly, let us introduce new variables r and θ such that

$$\begin{aligned} x_1 &= r\cos\theta \\ x_2 &= r\sin\theta \end{aligned} \tag{1.152}$$

In that case, the derivatives $\dot{x}_1$ and $\dot{x}_2$ become

$$\begin{aligned} \dot{x}_1 &= \dot{r}\cos\theta - (r\sin\theta)\dot{\theta} \\ \dot{x}_2 &= \dot{r}\sin\theta + (r\cos\theta)\dot{\theta} \end{aligned} \tag{1.153}$$

Substituting into (1.150), we can rewrite this equation as

$$\begin{aligned} \dot{r}\cos\theta - (r\sin\theta)\dot{\theta} &= p_1 r\cos\theta - p_2 r\sin\theta - r^3\cos\theta \\ \dot{r}\sin\theta + (r\cos\theta)\dot{\theta} &= p_2 r\cos\theta + p_1 r\sin\theta - r^3\sin\theta \end{aligned} \tag{1.154}$$

Sorting out the terms, we obtain

$$\begin{aligned} \cos\theta(\dot{r} - p_1 r + r^3) - \sin\theta(r\dot{\theta} - p_2 r) &= 0 \\ \sin\theta(\dot{r} - p_1 r + r^3) + \cos\theta(r\dot{\theta} - p_2 r) &= 0 \end{aligned} \tag{1.155}$$

Since (1.155) must equal zero for *all* θ, the following two conditions must hold

$$\begin{aligned} \dot{r} - p_1 r + r^3 &= 0 \\ r\dot{\theta} - p_2 r &= 0 \end{aligned} \quad \Rightarrow \quad \begin{aligned} \dot{r} &= r(p_1 - r^2) \\ r(\dot{\theta} - p_2) &= 0 \end{aligned} \tag{1.156}$$

Both equations are obviously satisfied for $r = 0$ (which corresponds to the equilibrium at the origin). However, when $p_1 > 0$, there is also a solution of the form

$$\begin{aligned} r(t) &= \sqrt{p_1} \\ \theta(t) &= p_2 t \end{aligned} \tag{1.157}$$

In the original coordinates, this solution can be expressed as

$$\begin{aligned} x_1(t) &= \sqrt{p_1}\cos p_2 t \\ x_2(t) &= \sqrt{p_1}\sin p_2 t \end{aligned} \tag{1.158}$$

This is clearly a limit cycle, which corresponds to a circle of radius $\sqrt{p_1}$ in the phase plane.

What can we conclude in this case?

1. $p_1 = 0$ is a bifurcation point, at which a limit cycle of amplitude $\sqrt{p_1}$ is born. In general, it can be shown that the amplitude of any limit cycle born through a Hopf bifurcation is always proportional to $(p - p_0)^{1/2}$, where p_0 is the bifurcation point.

2. It can be shown that the limit cycle in this example is an attractor for *all* values of $p_1 > 0$. An illustration of this fact is provided in Fig. 1.44, for $p_1 = 1$ and $p_2 = 1$.

3. A Hopf bifurcation which gives rise to an attractive limit cycle is called *supercritical*. If the limit cycle turns out to be unstable (and it can!), we refer to the bifurcation as *subcritical*.

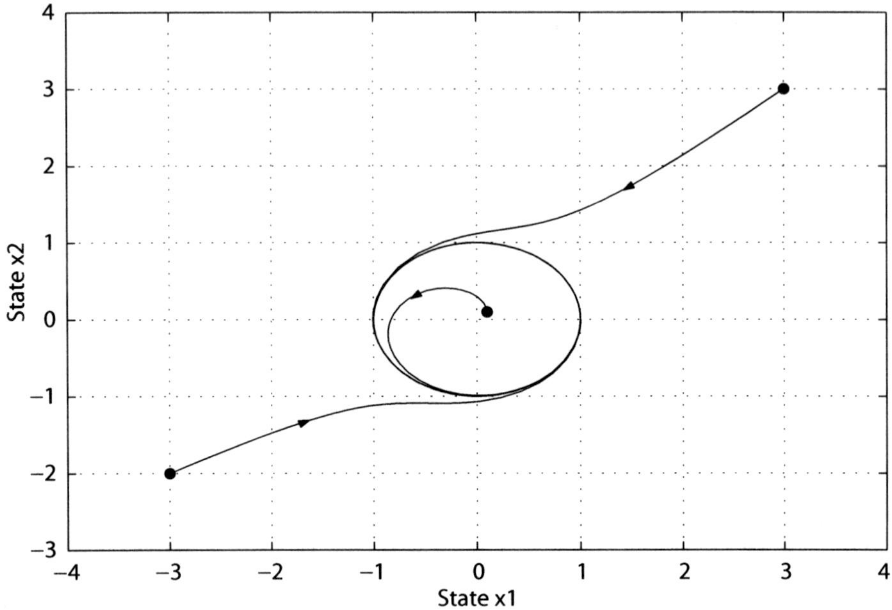

Fig. 1.44. The limit cycle for a Hopf bifurcation.

4. The notion of a Hopf bifurcation can be generalized to the scenario where an originally stable limit cycle loses this property, and in the process gives birth to a second limit cycle with a different frequency. In the following, we will refer to this process as a *generalized Hopf bifurcation.*

In systems of higher order, there are a number of other possibilities for bifurcations. Since a rigorous analysis of these phenomena is very complicated, in the following we will tend to describe them in terms of bifurcation diagrams. Before we apply this method, however, we need to address a conceptual problem that arises in this context. All our bifurcation diagrams up to this point involved plots of the equilibria as a function of the parameter. However, this approach doesn't work in the case of Hopf bifurcations, where we have to deal with a limit cycle that consists of infinitely many points. The problem is likely to get even worse for quasi-periodic and chaotic attractors. How can we incorporate different types of attractors into the bifurcation diagram without significantly complicating it? A simple way would be to plot only the *local maxima* of a trajectory that belongs to the attractor, and add them to the equilibrium points that are already in the diagram. In our example, *all* the local maxima of the limit cycle are equal to $\sqrt{p_1}$, so its contribution for a given value of p_1 would be a single point on the bifurcation diagram. The plot for system (1.150) obtained using this method is shown in Fig. 1.45.

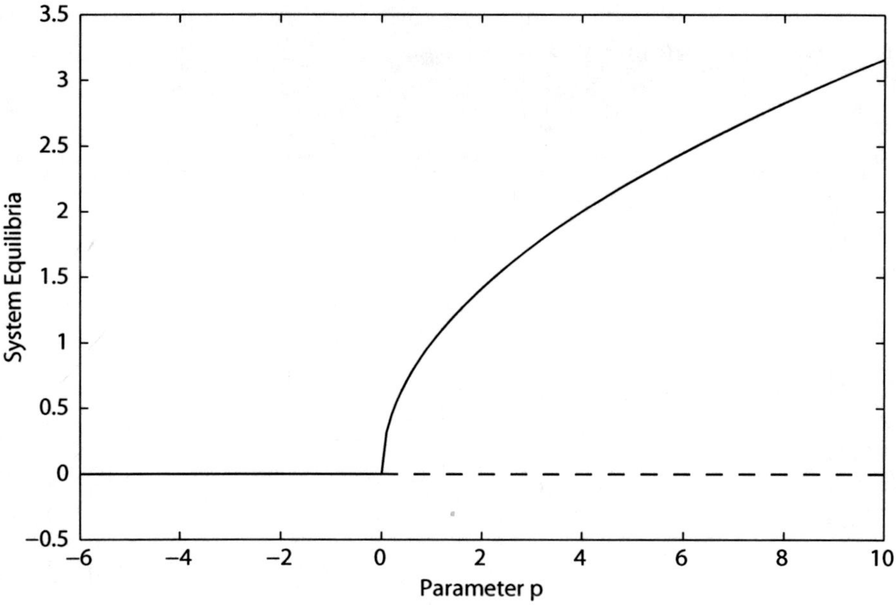

Fig. 1.45. The bifurcation diagram.

1.5 Routes to Chaos

In this section we will consider three different mechanisms through which chaos can arise. In all cases, we will observe how the system behavior changes as one or more parameters are varied. Bifurcation diagrams will prove to be a very useful tool for this purpose.

The Period-Doubling Route

Consider a system with a stable limit cycle of period T. For a given parameter value p, the amplitude of the periodic solution can be represented as the uniform sequence shown in Fig. 1.46. In the bifurcation diagram, this sequence would obviously correspond to a single point, since all maxima are equal to 1.

As the parameter is increased beyond some value p_1, the situation shown in Fig. 1.47 can arise. The new waveform is still periodic, but its period is now $2T$, due to the variations in amplitude. Since the steady state solution now has two different peak values, the bifurcation diagram 'splits' in the manner shown in Fig. 1.48.

The 'period-doubling' bifurcation described above can occur again for some $p = p_2$, at which point the peak values in the system have the form shown in Fig. 1.49. The period becomes $4T$, and the bifurcation diagram splits into four branches, as in Fig. 1.50.

It is well known that a cascade of period-doubling bifurcations can ultimately lead to chaos. This phenomenon has been observed in a wide variety of models, including several that we already considered. A typical example of this phenomenon is shown in Fig. 1.51.

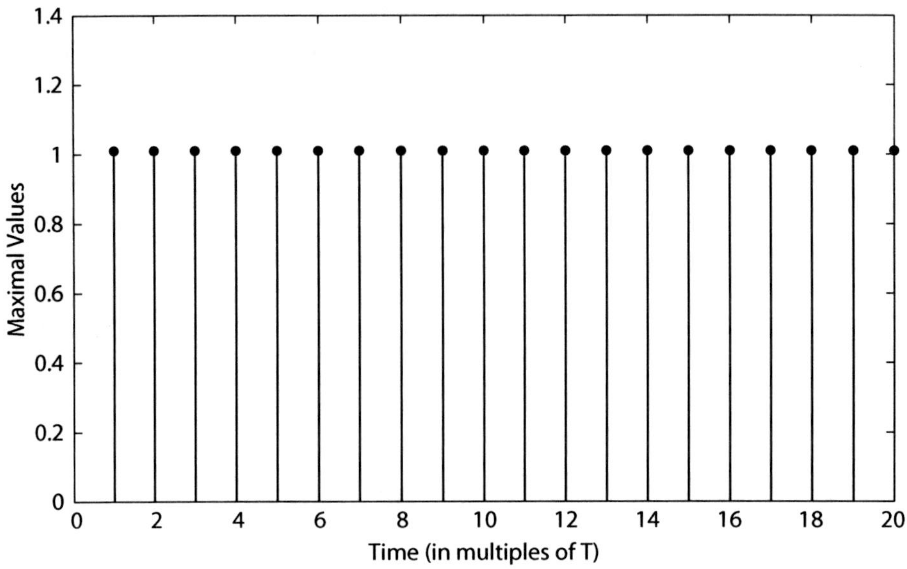

Fig. 1.46. Peak values for a limit cycle with a single amplitude.

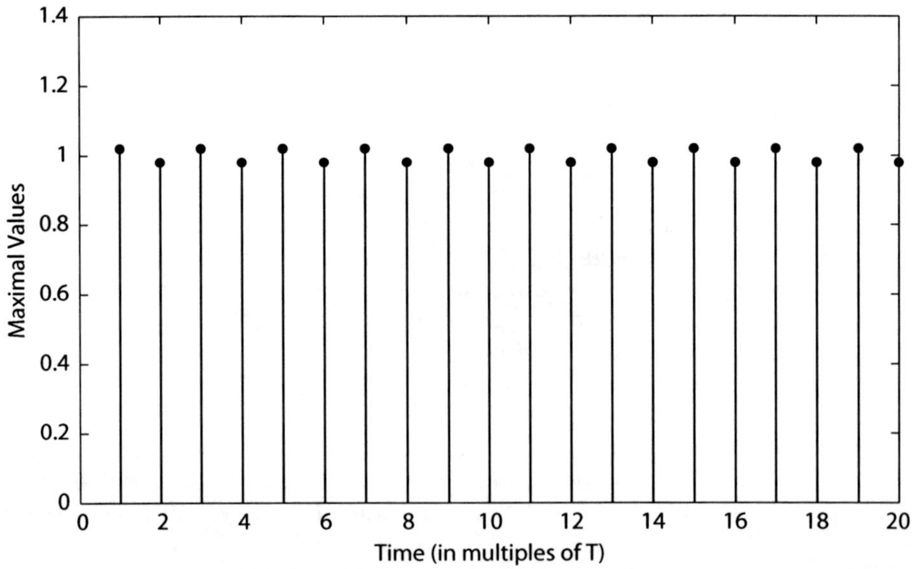

Fig. 1.47. Peak values for a limit cycle with two amplitudes.

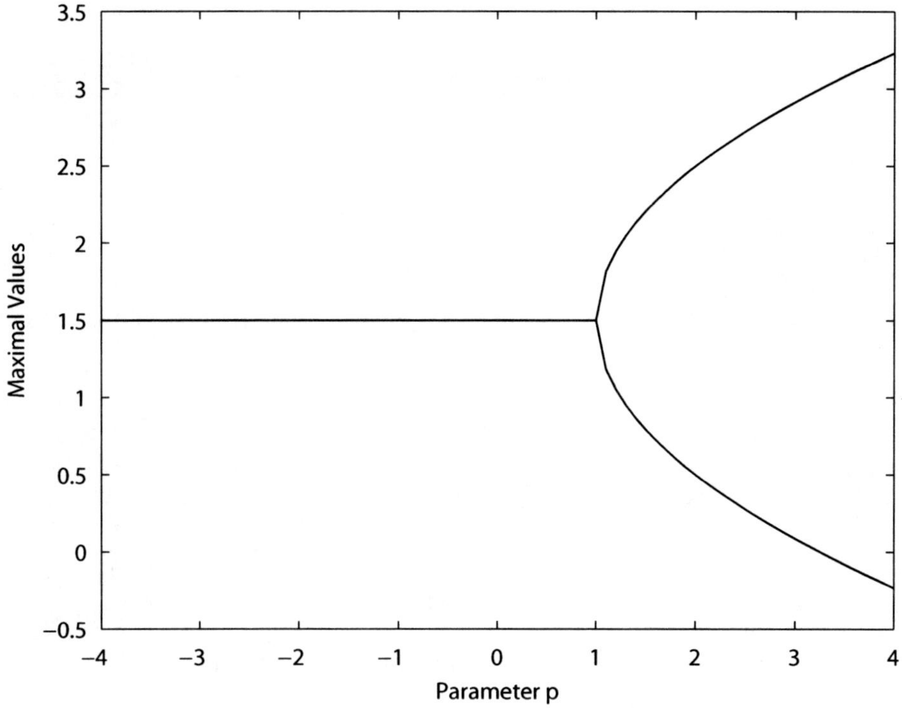

Fig. 1.48. Splitting of the bifurcation diagram.

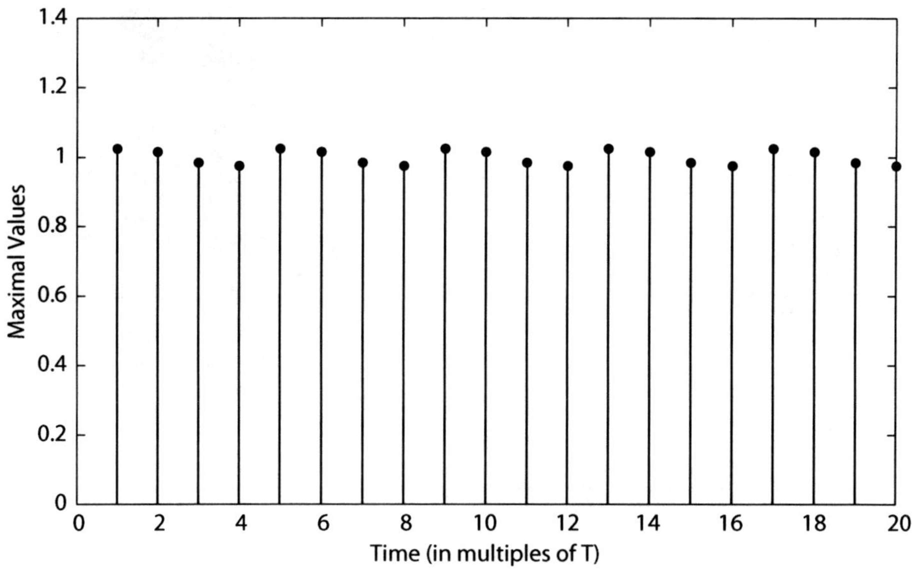

Fig. 1.49. Peak values for a limit cycle with four amplitudes.

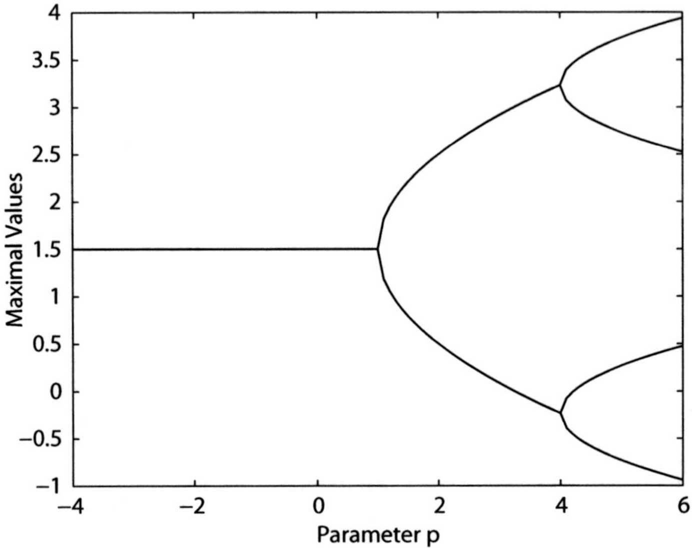

Fig. 1.50. Bifurcation diagram with two splits.

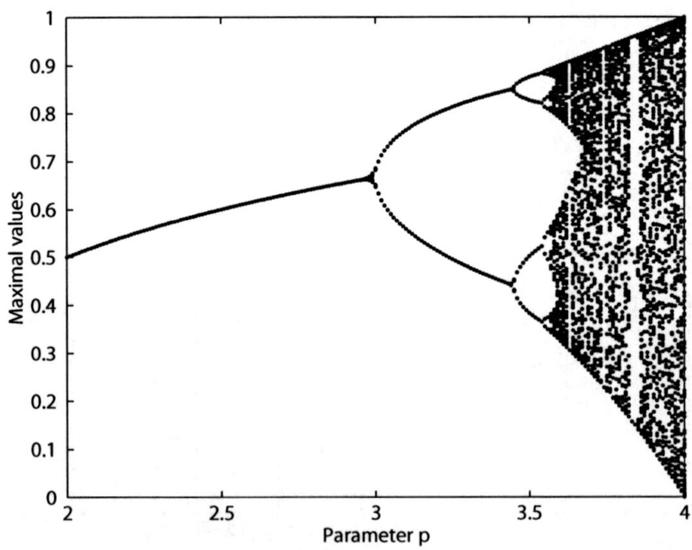

Fig. 1.51. The bifurcation diagram for a chaotic system.

A rigorous examination of conditions under which period-doubling occurs is not possible at this time, since the underlying math is far too complicated. Nevertheless, a simple (and admittedly crude) one dimensional model can provide considerable insight into what is really going on. To that effect, let us assume that any two successive maxima of our trajectory are related as

$$x_{k+1} = f(x_k) = px_k(1 - x_k) \tag{1.159}$$

How one can come up with such an equation in general remains something of a mystery. What is important, however, is that this is a simple scalar *difference equation* which accounts for parameter variations, and allows us to easily plot the maxima as a function of p. To illustrate this point, let us set $p = 2$ and pick $x_1 = 0.2$ as the initial point. Following (1.159), we easily obtain the sequence shown in Fig. 1.52, which clearly indicates that there is a *single* steady state amplitude $x^* = 0.5$.

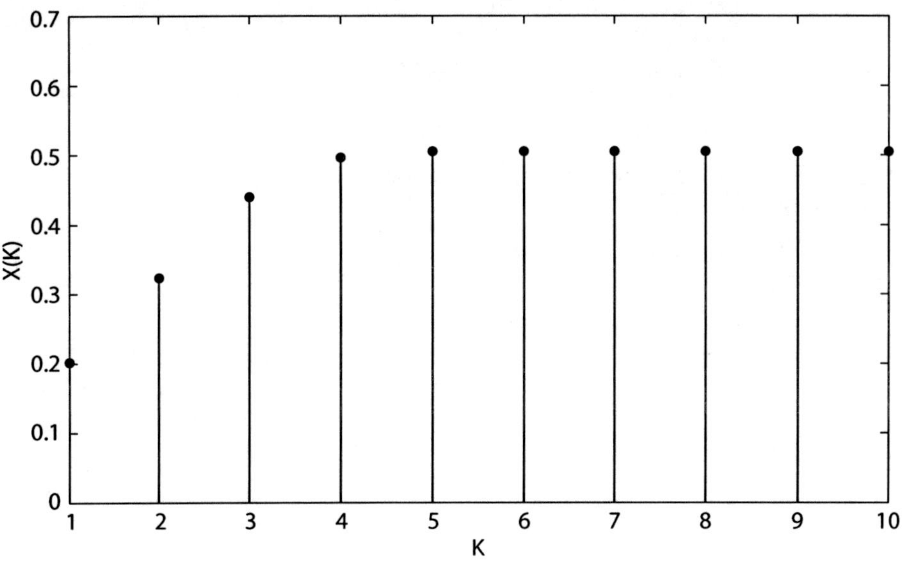

Fig. 1.52. The sequence for $x_1 = 0.2$.

A more intuitive way of looking at equation (1.159) would be to generate the sequence $\{x_k\}$ in the manner shown in Fig. 1.53. In this figure we plotted two separate functions: $x_{k+1} = px_k(1 - x_k)$ and $x_{k+1} = x_k$. The intersection points represent the equilibria of system (1.159), since they satisfy the condition

$$x^* = f(x^*) \tag{1.160}$$

It is not difficult to show that equilibrium $x^* = 0$ is unstable, while the one corresponding to $x^* = 0.5$ is stable. Indeed, a simple inspection of Figs. 1.54 and 1.55 shows that the stability of the equilibrium depends exclusively on the slope of f at the intersection point. If the magnitude of slope is < 1 (as in Fig. 1.54), the equilibrium is stable; if not, the intersection point is a repeller.

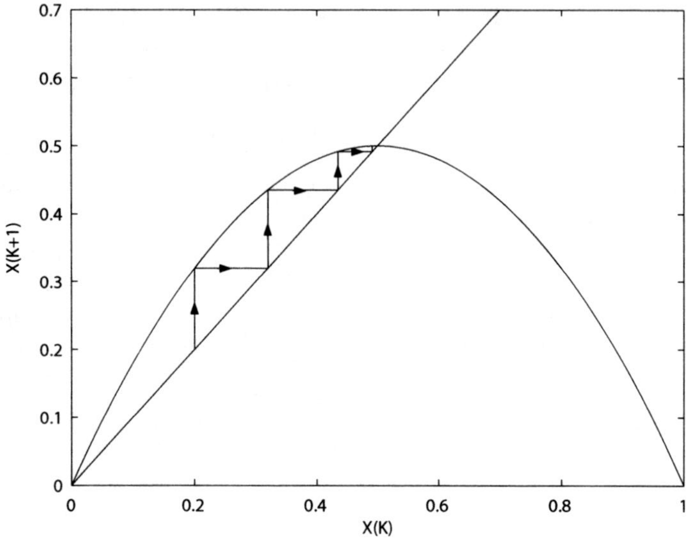

Fig. 1.53. A schematic approach for generating the sequence.

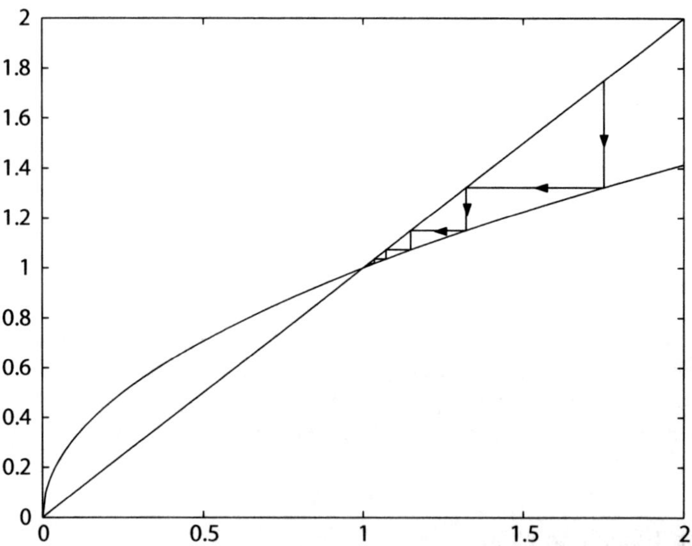

Fig. 1.54. The sequence for a slope < 1.

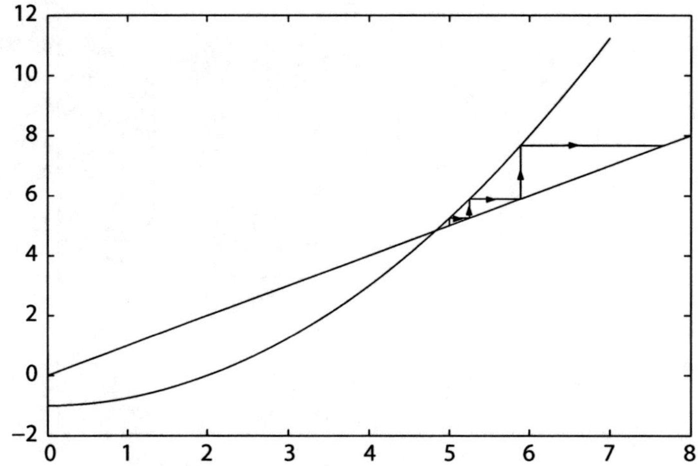

Fig. 1.55. The sequence for a slope > 1.

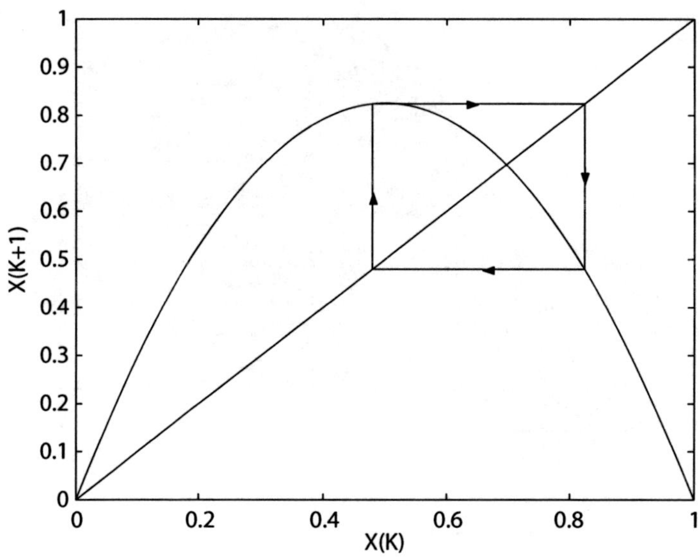

Fig. 1.56. A schematic representation of the two-amplitude case.

Let us now see what happens when p is increased to $p = 3.3$. In this case, the graphical method gives rise to the situation shown in Fig. 1.56, which suggests that there are *two* alternating maxima in the steady state. This is consistent with the amplitude splitting model shown in Fig. 1.48, and indicates that a period-doubling bifurcation has occurred.

A further illustration of this phenomenon is provided in Fig. 1.57, for $p = 3.53$. In this case, the solution has period $4T$, with *four* distinct maxima.

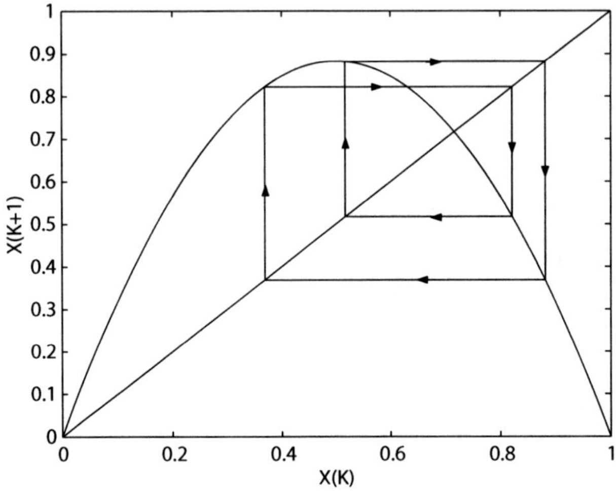

Fig. 1.57. A schematic representation of the four-amplitude case.

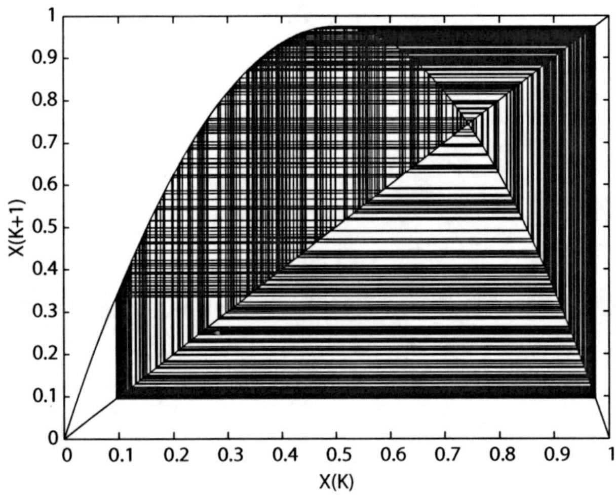

Fig. 1.58. A schematic representation of the chaotic case.

To verify that this mechanism leads to chaos, let us consider system (1.159) when p is increased to $p = 3.9$. The situation that corresponds to this case is shown in Fig. 1.58, which has all the characteristics of chaotic behavior (no periodicity or convergence, bounded maxima).

The Quasi-Periodic Route

Suppose that the system has a stable limit cycle of period T for parameter values satisfying $p \leq p_0$. As noted earlier, a *generalized* Hopf bifurcation associated with this cycle can give birth to a *second* limit cycle, whose frequency is $\omega_2 \neq \omega_1$.

If the two frequencies turn out to be incommensurate, the attractor becomes quasi-periodic with two independent frequencies. This type of Hopf bifurcation can reoccur for some $p = p_1$, giving rise to a *third* limit cycle of period ω_3. If ω_3 is incommensurate with ω_2 and ω_1, we ultimately arrive at a quasi-periodic attractor consisting of three frequencies.

It turns out that the stability of such a system can be very fragile, and that chaotic behavior can be triggered by a small perturbation. This sequence of events was theoretically predicted by Ruelle and Takens in 1971,[6] and was subsequently observed experimentally both in fluid dynamics and in electronic circuits.[7]

The Intermittent Route

The third route to chaos is rather direct, and doesn't require multiple bifurcations. To see why this route is associated with 'intermittency', let us assume that there is a limit cycle of period T for parameter values $p \leq p_1$. As we pass p_1, there will be sudden short bursts of aperiodic behavior followed by relatively long periods of regular oscillations (the oscillations are not strictly periodic, but are very close). This type of behavior is illustrated in Fig. 1.59. Note that as p moves further away form p_1, the intermittent chaotic bursts become longer and longer, until all traces of periodic behavior vanish.

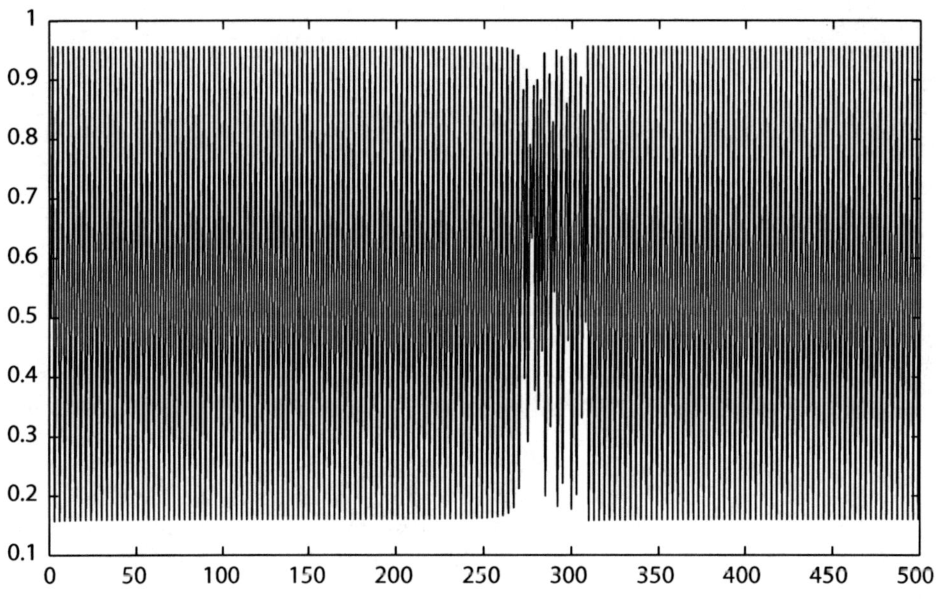

Fig. 1.59. An illustration of intermittent behavior.

The intermittent route to chaos has a nice interpretation in terms of discrete one dimensional models. To see that, let us assume once again that successive maxima of the trajectory satisfy equation (1.159). However, this time we will not plot x_{k+1} versus x_k, but rather x_{k+3} (that is, every *third* maximum). Such a plot is easily generated from (1.159), and is shown in Fig. 1.60 for $p = 3.87$.

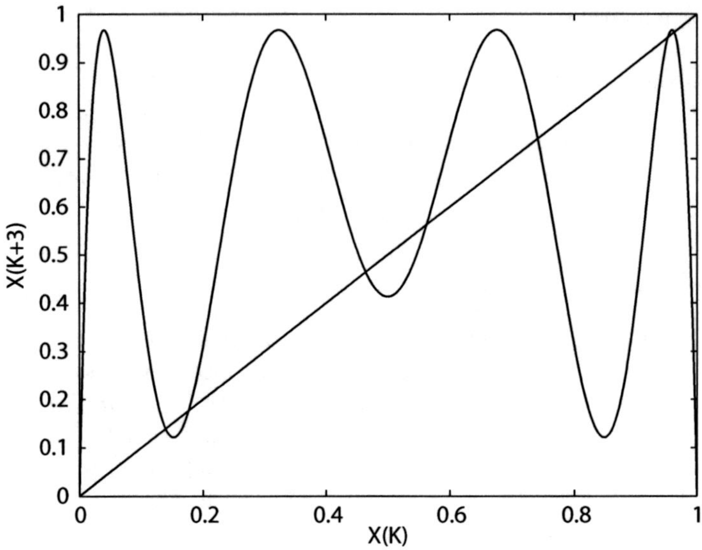

Fig. 1.60. The schematic representation for $p = 3.87$.

The plot in Fig. 1.60 indicates six points of intersection, three of which have slopes < 1 in magnitude (and are therefore attractors). Since we are plotting x_{k+3} versus x_k, each of the three fixed points corresponds to a different amplitude (the values are $x^* = 0.16$, $x^* = 0.51$, and $x^* = 0.95$, respectively).

Let us now consider a slightly smaller value of p, such as $p = 3.82$. The plot that corresponds to this case is shown in Fig. 1.61, and a closer inspection reveals that there are no more stable fixed points (the only remaining intersections have slopes > 1).

This obviously means that no periodic solutions exist. However, if we magnify the section of the curve close to the straight line (as shown in Fig. 1.62), we notice that x_k can spend a considerable amount of time in the 'gap' where $x_{k+3} \approx x_k$. During such intervals, the solution is *almost* periodic. Once the trajectory leaves the gap, it roams around freely until it becomes trapped in the gap again. Based on this model, it is natural to associate the 'free roaming' intervals with the short aperiodic bursts, and the intervals corresponding to the gap with almost periodic behavior. Of course, as the curve moves away from the line, x_k spends less and less time in the gap, until the solution becomes completely aperiodic.

1.6 Chaos in Physical Systems

Chaotic behavior has been observed in a variety of engineering models, ranging from classical Newtonian mechanics to fluid dynamics and electronic circuits.

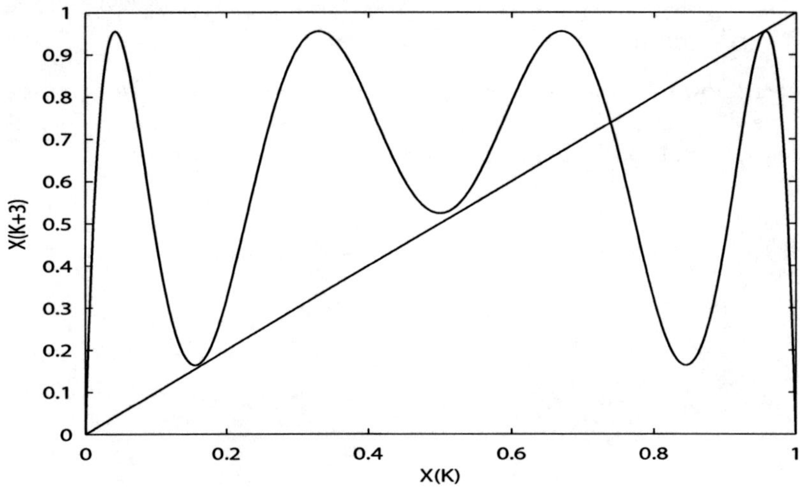

Fig. 1.61. The schematic representation for $p = 3.82$.

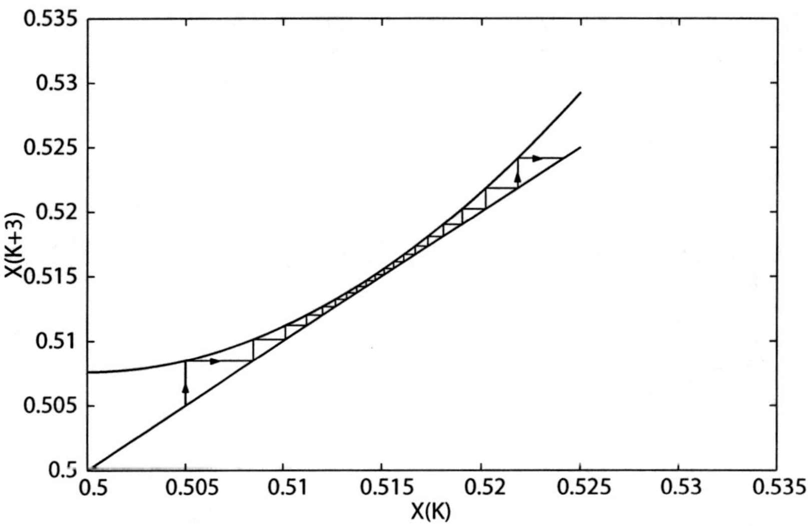

Fig. 1.62. A schematic explanation for intermittent aperiodic behavior.

In the following, we will consider two simple physical systems that exhibit this property.

The Forced Pendulum

As is well known, the motion of a pendulum is governed by Newton's Second Law. The diagram of forces is shown in Fig. 1.63, with θ and g denoting the displacement angle and the gravitational acceleration, respectively.

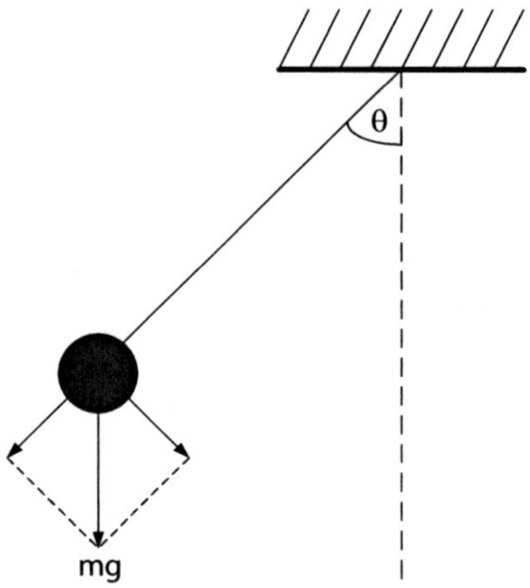

Fig. 1.63. The forces acting on the pendulum.

It is usually assumed that the damping force is proportional to the angular velocity, so the overall equation of motion can be expressed as

$$m\ddot{x} = -\gamma\dot{\theta} - mg\sin\theta \tag{1.161}$$

Recalling that

$$dx = ld\theta \tag{1.162}$$

it is easily verified that (1.161) can be rewritten as a second order differential equation in $\theta(t)$

$$ml\ddot{\theta} + \gamma\dot{\theta} + mg\sin\theta = 0. \tag{1.163}$$

The dynamics of the pendulum become considerably more interesting if we include an external sinusoidal force

$$F_e = A\cos\omega_D t \tag{1.164}$$

with a *fixed frequency* ω_D and a *variable amplitude* A. In that case, the model becomes

$$\ddot{\theta} + a_1\dot{\theta} + a_2\sin\theta = p\cos\omega_D t \tag{1.165}$$

where $a_1 = \gamma/ml$ and $a_2 = g/l$ are constants, and $p = A/ml$ is a parameter. Defining $\omega = \dot{\theta}$ and setting $a_2 = 1$ (which can be done by choosing l to equal the gravitational acceleration), we can reformulate (1.165) in the form of state equations

$$\begin{aligned} \dot{\theta} &= \omega \\ \dot{\omega} = \ddot{\theta} &= -a_1\omega - \sin\theta + p\cos\omega_D t \end{aligned} \tag{1.166}$$

Since this is a *non-autonomous* system, it is useful to define an additional state variable as $\phi = \omega_D t$. In that case, we obtain a third-order model

$$\begin{aligned} \dot{\theta} &= \omega \\ \dot{\omega} &= -a_1\omega - \sin\theta + p\cos\phi \\ \dot{\phi} &= \omega_D \end{aligned} \tag{1.167}$$

which contains no explicit dependency on time.

To see how chaotic behavior can arise in this system, let us set $a_1 = 0.5$ and $\omega_D = 2/3$, and allow parameter p to vary. The system dynamics for two different values of p are shown in Figs. 1.64 - 1.67, which indicate that both periodic and chaotic motion is possible for a forced pendulum.

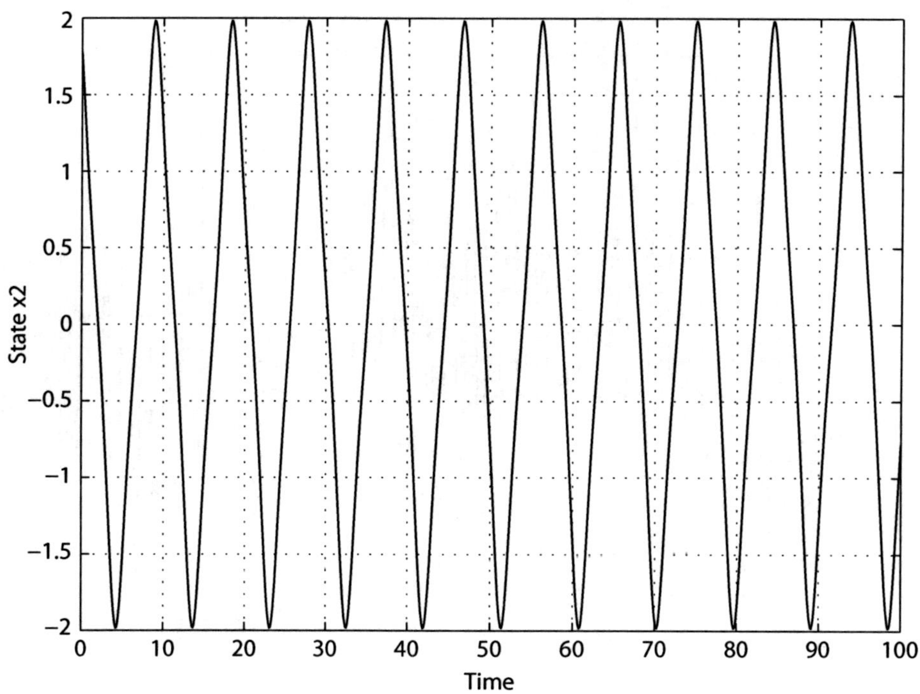

Fig. 1.64. The periodic motion of a forced pendulum.

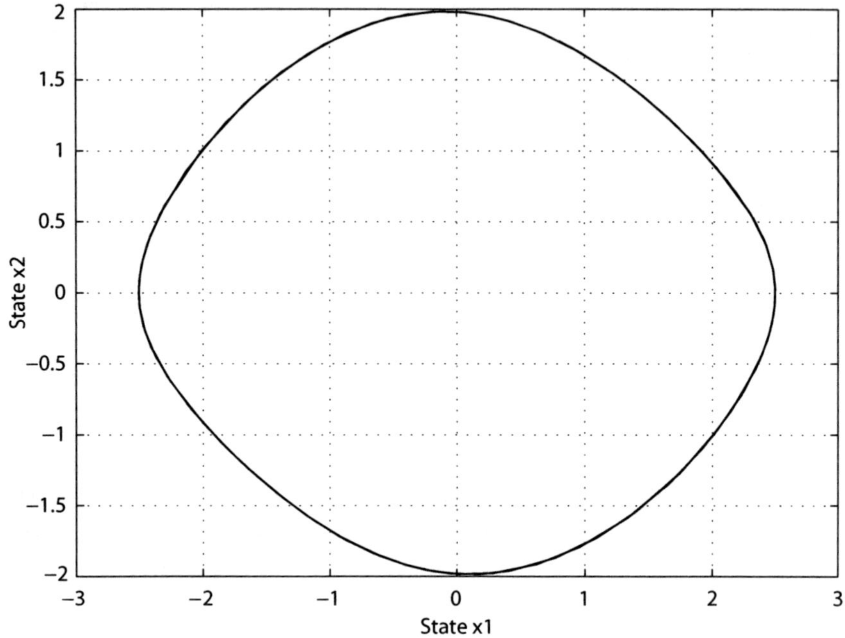

Fig. 1.65. The limit cycle for the periodic motion.

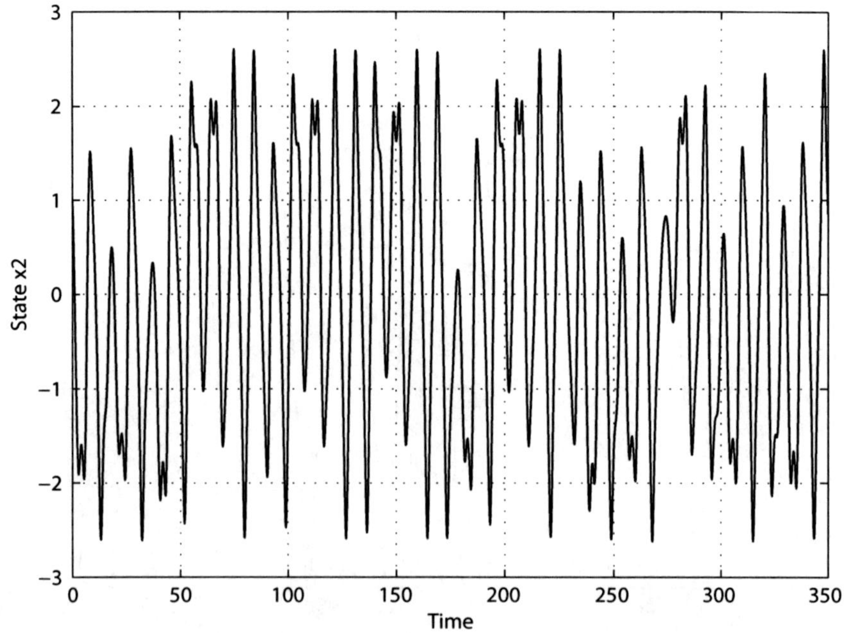

Fig. 1.66. The chaotic motion of a forced pendulum.

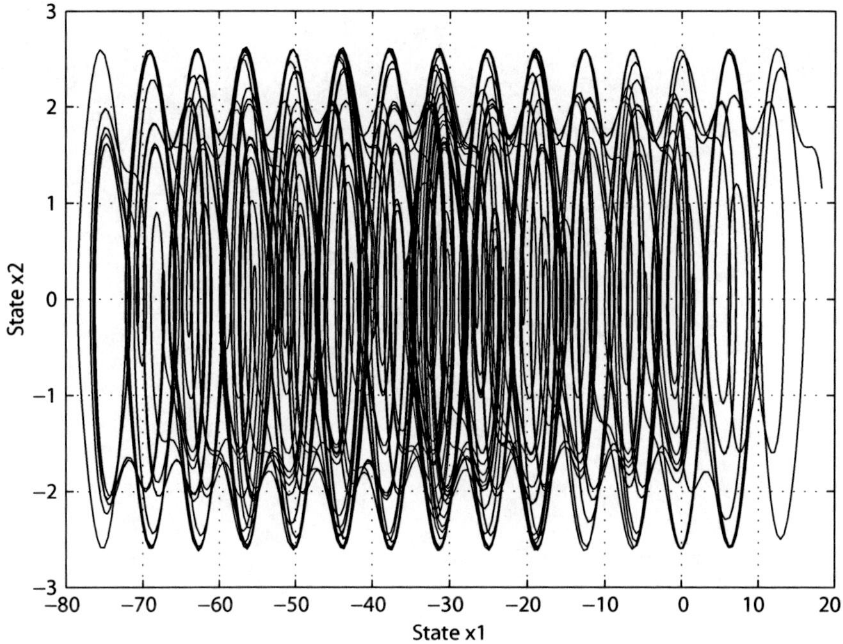

Fig. 1.67. The strange attractor for the chaotic motion.

Chua's Circuit

The circuit shown in Fig. 1.68 was first considered by L. O. Chua.

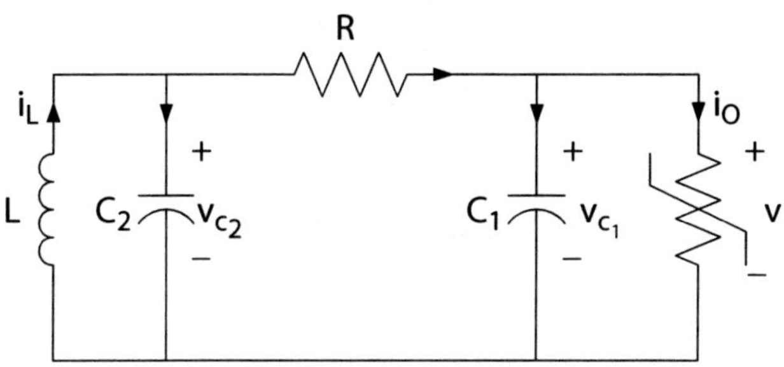

Fig. 1.68. Chua's nonlinear circuit.

The nonlinear resistor is defined by the piecewise-linear current-voltage relationship

$$f(v) = \begin{cases} m_0 v + B_p(m_1 - m_0), & v \geq B_p \\ m_1 v, & -B_p < v < B_p \\ m_0 v - B_p(m_1 - m_0), & v \leq -B_p \end{cases} \tag{1.168}$$

whose graphical representation is provided in Fig. 1.69.

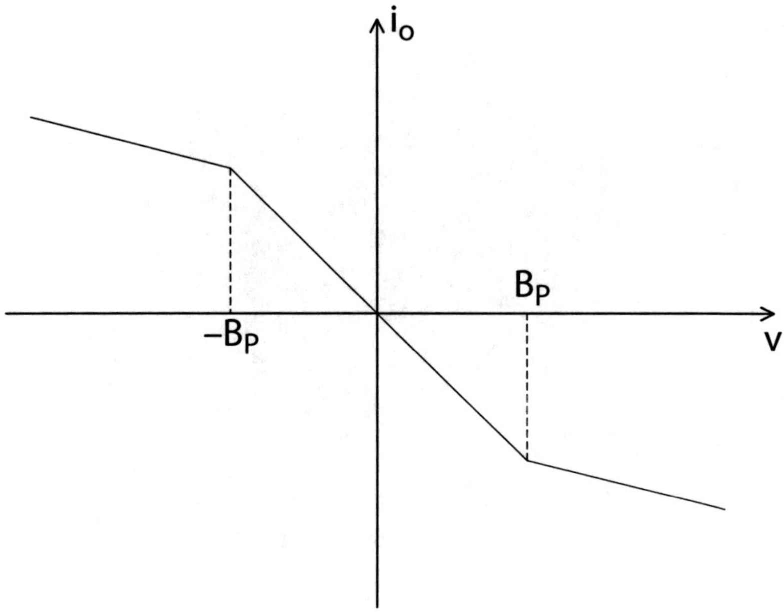

Fig. 1.69. The current-voltage characteristic of the nonlinear element.

The state model for this circuit can be derived from the following four equations, which correspond to Kirchoff's laws and Ohm's law

KCL	KVL	Ohm's Law	
$-i_L + i_{C2} + i_R = 0$	$v_L + v_{C2} = 0$	$i_R = G(v_{C2} - v_{C1})$	(1.169)
$-i_R + i_{C1} + i_0 = 0$			

Recalling that $i_0 = f(v_{C1})$, equations (1.169) yield

$$\begin{aligned} i_{C2} &= C_2\dot{v}_{C2} = i_L - G(v_{C2} - v_{C1}) \\ i_{C1} &= C_1\dot{v}_{C1} = G(v_{C2} - v_{C1}) - f(v_{C1}) \\ v_L &= L\dot{i}_L = -v_{C2} \end{aligned} \tag{1.170}$$

In order to simplify the analysis, let us additionally assume that $G = C_2$ (numerically, not dimensionally!) and define the following scaled variables

$$x_1 = \frac{v_{C1}}{B_p}, \qquad x_2 = \frac{v_{C2}}{B_p}, \qquad x_3 = \frac{1}{B_p G} i_L \tag{1.171}$$

It is easily verified that the system can now be described as

$$\begin{aligned}\dot{x}_1 &= \alpha[-x_1 + x_2 - h(x_1)] \\ \dot{x}_2 &= x_1 - x_2 + x_3 \\ \dot{x}_3 &= -\beta x_2\end{aligned} \tag{1.172}$$

where

$$\alpha = \frac{G}{C_1} = \frac{C_2}{C_1}, \qquad \beta = \frac{1}{LG} \tag{1.173}$$

and

$$h(x_1) = \frac{1}{B_p G} f(v_{C1}) = \begin{cases} bx_1 + (a-b), & x_1 \geq 1 \\ ax_1, & -1 < x_1 < 1 \\ bx_1 - (a-b), & x_1 \leq -1 \end{cases} \tag{1.174}$$

with $a = m_1/G$ and $b = m_0/G$.

It should be noted that (1.172) resembles the double-scroll model (1.114), the only difference being the addition of term $x_1(t)$ in the first equation. It is therefore not surprising that this circuit exhibits chaotic behavior for certain parameter values. The dynamics of this system are illustrated in Figs. 1.70 - 1.73, which correspond to two different values of parameter α (with a, b and β set to $a = -8/7$, $b = -5/7$ and $\beta = 16$ in both cases).

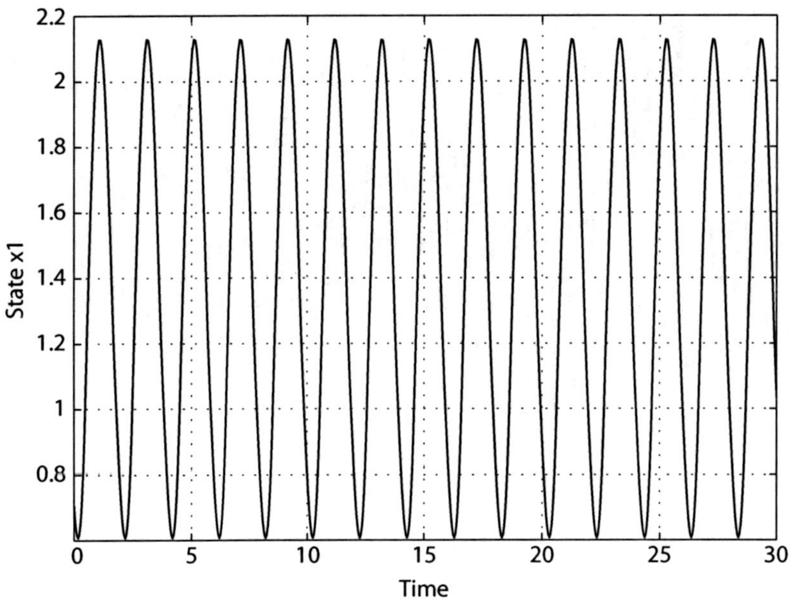

Fig. 1.70. A periodic solution for Chua's circuit.

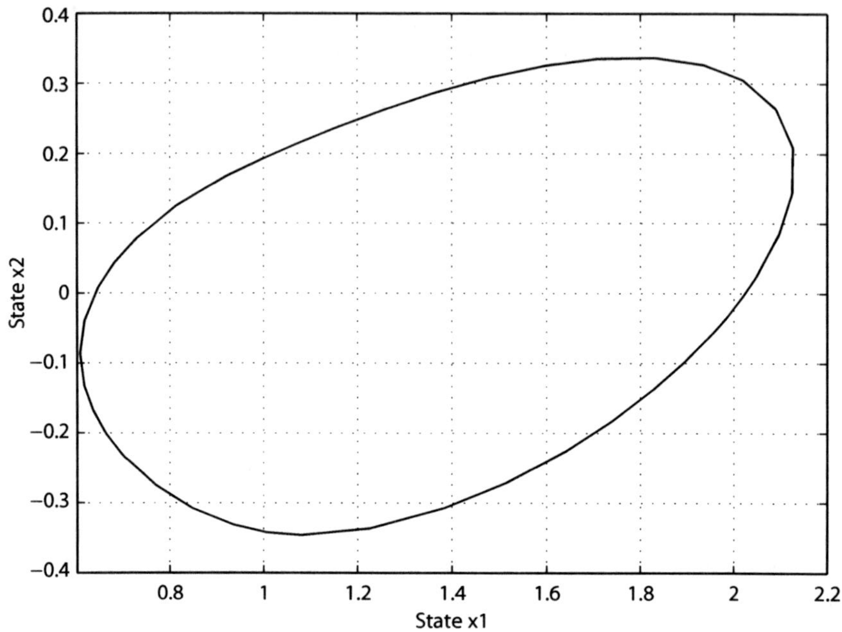

Fig. 1.71. The limit cycle corresponding to the periodic solution.

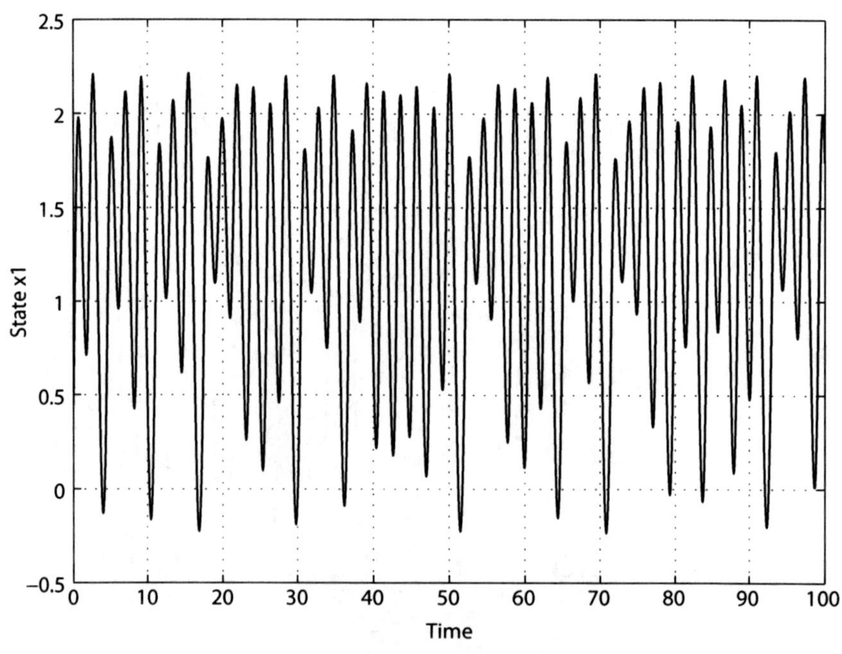

Fig. 1.72. Chaotic oscillations in Chua's circuit.

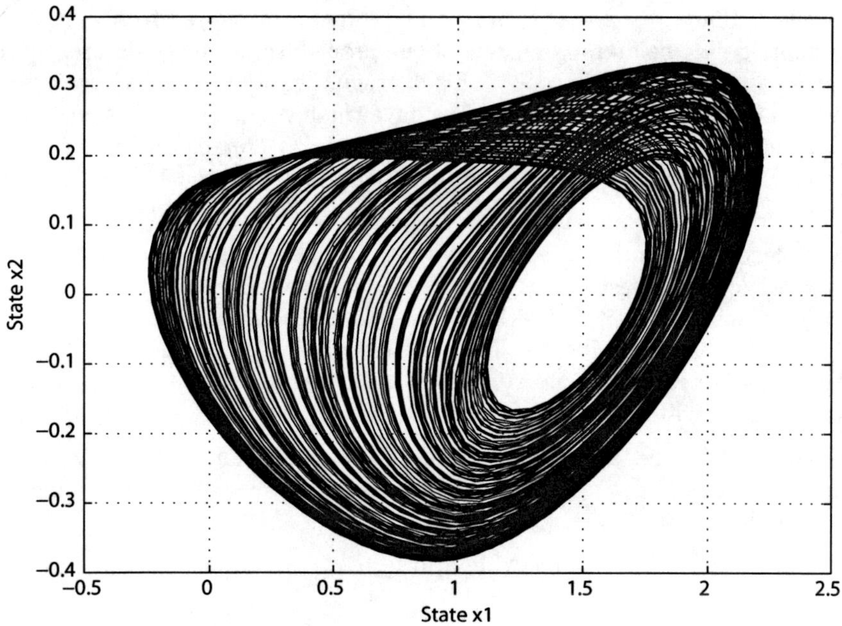

Fig. 1.73. The strange attractor for the chaotic case.

Attractor Reconstruction

The previous two examples have well defined analytical models, and the corresponding attractors can be constructed quite accurately using numerical simulation. But what if we are dealing with a physical system whose model is *not* known, and for which we can measure only a single output variable $y(t)$? Is it possible to detect chaos in such a system, and perhaps distinguish it from purely random behavior? Amazingly, in some cases it is. It turns out that if the system has an attractor of low dimension, we can approximately "reconstruct" this set.

In order to understand how this can be done, we first need to revisit the problem of estimating fractal dimensions. In the case of chaotic attractors, the capacitive dimension introduced in Section 1.3 is usually a rather impractical measure, since it generally requires a great deal of computer time and memory. An alternative definition with better numerical properties is the so-called *correlation dimension*, whose calculation is explained below.

STEP 1. Generate a large set of points on the attractor (a single solution will do if we run it long enough to eliminate all transients).

STEP 2. Fix a point x on the attractor, and let $N_x(\varepsilon)$ denote the number of points from this set that are contained in an n-dimensional ball of radius ε centered at x.

STEP 3. As ε increases, $N_x(\varepsilon)$ is expected to grow as $N_x(\varepsilon) \sim \varepsilon^{d_x}$. The exponent d_x that characterizes this function is referred to as the *pointwise dimension* at x.

STEP 4. Since x is an arbitrary point (and can be chosen from a sparser or denser portion of the attractor), repeat the procedure for many different points, and average $N_x(\varepsilon)$ for a fixed ε. This average, denoted $C(\varepsilon)$, is considered to be independent of x, and is assumed to have the form $C(\varepsilon) \sim \varepsilon^d$. The exponent d in this expression is referred to as the *correlation dimension.* Typically, this dimension is somewhat smaller than its capacitive counterpart, but the two are usually similar.

In order to determine d explicitly, it usually suffices to plot $\log C(\varepsilon)$ as a function of $\log \varepsilon$. In estimating the slope, one should bear in mind that the graph is linear only in its "middle" portion. Indeed, for large values of ε, the n-dimensional ball includes the *entire* attractor, so $N_x(\varepsilon)$ cannot increase any further (which creates a saturation effect). On the other hand, when ε is sufficiently small, x will be the only point that is included in $N_x(\varepsilon)$.

Having defined an appropriate measure of dimensionality, we can now describe the process of attractor reconstruction in terms of the following three step algorithm.

STEP 1. Fix a delay value Δ, and form a $k+1$ - dimensional vector

$$z(t) = \begin{bmatrix} z_1(t) \\ z_2(t) \\ z_3(t) \\ \vdots \\ z_{k+1}(t) \end{bmatrix} = \begin{bmatrix} y(t) \\ y(t-\Delta) \\ y(t-2\Delta) \\ \vdots \\ y(t-k\Delta) \end{bmatrix} \tag{1.175}$$

Then, plot the evolution of z in the $k+1$ - dimensional state space, and identify the attractor (if one exists).

STEP 2. Estimate the correlation dimension of the attractor, using the procedure described above.

STEP 3. If the correlation dimension is larger than in the previous step, increment k and repeat the process. If it remains unchanged, stop.[8]

It has actually been shown that when the system has a low order chaotic attractor, then there exists a delay Δ and an "embedding dimension" k which allow for its successful reconstruction.[9] The problem with this result is that it is not clear how to optimally choose Δ, so in most cases the appropriate delay needs to be determined heuristically. Despite this weakness, however, it is fair to say that the possibility of attractor reconstruction remains a truly remarkable result. Edward Lorenz (the scientist who discovered chaos) actually claimed that this was the most surprising property that he encountered in his study of nonlinear systems.[10]

1.7 Notes

1. E. N. Lorenz, "Deterministic non-periodic flow," *Journal of Atmospheric Science*, **20**, pp. 130-141, 1963.

2. O. E. Rössler, "An equation for continuous chaos," *Physics Letters A*, **57**, pp. 397-398, 1976.

3. T. Matsumoto, "A chaotic attractor from Chua's circuit," *IEEE Transactions on Circuits and Systems*, **CAS-31**, pp. 1055-1058, 1984.

4. Y. Ueda, "Survey of regular and chaotic phenomena in the forced Duffing oscillator," *Chaos, Solitons and Fractals*, **1**, pp. 199-231, 1991.

5. It should be noted that the difference $\delta x(t)$ starts out as very small, and remains that way for a certain amount of time by virtue of (1.118).

6. D. Ruelle and T. Takens, "On the nature of turbulence," *Communications in Mathematical Physics*, **20**, pp. 167-192, 1971.

7. P. Linsay and A. Cumming, "Three-frequency quasiperiodicity, phase locking and the onset of chaos," *Physica D*, **40**, pp. 196-217, 1989.

8. If the system has a low order chaotic attractor, the correlation dimension computed by this algorithm should converge to a small number. This is quite different from purely random systems, where the dimension increases monotonically with k.

9. See, for example: Floris Takens, "Detecting strange attractors in turbulence," in *Dynamic Systems and Turbulence - Warwick 1980*, D. A. Rand and L. S. Young (Eds.), Springer-Verlag, 1981.

10. S. Strogatz, *Nonlinear Dynamics and Chaos*, Perseus Publishing, 1994.

Further Reading on Chaos Theory

The books suggested in the following are just a minimal sample of the existing literature on chaos theory. A comprehensive bibliography with over 7,000 items can be found in:

S. Y. Zhang, *Bibliography on Chaos*, World Scientific, 1991.

Introductory Books

E. N. Lorenz, *The Essence of Chaos*, University of Washington Press, 1999.

I. Prigogine and I. Stengers, *Order Out of Chaos*, Bantam Books, 1984.

I. Stewart, *Does God Play Dice?*, Blackwell, 1989.

J. Gleick, *Chaos: Making a New Science*, Penguin Group, 1988.

Textbooks and Monographs

T. S. Parker and L. O. Chua, *Practical Numerical Algorithms for Chaotic Systems*, Springer Verlag, 1989.

T. Kapitniak, *Chaos for Engineers: Theory, Applications and Control*, Springer Verlag, 2000.

R. C. Hilborn, *Chaos and Nonlinear Dynamics*, Oxford University Press, 1994.

G. Baker and J. Gollub, *Chaotic Dynamics: An Introduction*, Cambridge University Press, 1990.

S. Strogatz, *Nonlinear Dynamics and Chaos*, Perseus Publishing, 1994.

Chaos and Theology

R. J. Russell, N. Murphy and A. R. Peacocke (Eds.), *Chaos and Complexity: Scientific Perspectives on Divine Action*, Vatican Observatory Publications, 2000.

D. Mackay, *Science, Chance and Providence*, Oxford University Press, 1978.

S. Bonting, *Chaos Theology: A Revised Creation Theology*, Novalis, 2002.

J. Polkinghorne, *Quarks, Chaos and Christianity*, Crossroad, 2000.

Chapter 2

Metamathematics

Metamathematics, as the name suggests, goes beyond the derivation of new results from existing axioms and theorems, and focuses instead on the *structure* of proofs. To understand why this sort of analysis is important, we must first say a few words about axioms and the way in which they are chosen. The ancient Greeks believed that axioms ought to be self-evident statements which are consistent with our everyday experience, and can therefore be accepted without proof. This conviction lies at the heart of Euclidean geometry, which grew out of five seemingly obvious truths about lines, angles and circles.

For many centuries, this discipline was viewed as the cornerstone of mathematics, whose logical foundations were beyond doubt. There was, however, a nagging problem with Euclid's fifth axiom, which continued to trouble generations of theoreticians. This axiom (which is known as the Parallel Postulate) can be paraphrased as follows:

PROPOSITION 5. Consider a line l and a point a that does not belong to it. Then, there is one and only one line that passes through a and does not intersect l.

The difficulty with the Parallel Postulate stems from the fact that it deals with lines of infinite length. The mathematicians of ancient Greece (and those who followed them in subsequent centuries) could not accept such a statement as a "self-evident" truth, since they believed that human intuition is limited to finite objects. As a result, the two millennia that followed Euclid's death witnessed countless attempts to close the loophole, and derive Proposition 5 from the other four axioms. The status of this proposition was finally resolved in the 19th century, when mathematicians Nikolai Lobachevsky and János Bolyai established that such a derivation is impossible.

The recognition that Proposition 5 was *unprovable* in the framework of classical geometry formally justified Euclid's decision to incorporate it into his system as an additional axiom. However, it also opened up the possibility that the *opposite* of this proposition could be an axiom in its own right (provided, of course, that the resulting system remains free of contradictions). This realization led to the development of several non-Euclidean geometries, whose results are anything but "self-evident." Although such geometries seem to defy our intuitive understanding of spatial relationships, it would be a mistake to dismiss them as "artificial" constructs of the human mind. Einstein's theory of general relativity provides an excellent illustration of this point, since it uses non-Euclidean geometry to describe the structure of space and time (both of which are fundamental constituents of physical reality).

One of the most important insights that can be gained from the results of Lobachevsky and Bolyai is that axioms can actually be completely divorced from human experience. Nineteenth century mathematicians quickly capitalized on this possibility, and began to think of axioms as strings of completely meaningless symbols which can be manipulated according to a set of precisely defined rules. This conceptual shift marked the birth of a new paradigm, whose central claim was that *all* mathematical thinking could be formally described in such abstract terms.

Throughout the early decades of the 20th century, it was widely believed that this approach would eventually resolve all remaining ambiguities, and that the Platonic ideal of a "perfect" mathematical theory was finally within our reach. This seemingly unstoppable tide of optimism was suddenly reversed in 1931, when Austrian mathematician Kurt Gödel published his famous Incompleteness Theorem. Gödel showed that any sufficiently complex formal system must necessarily contain unprovable propositions, and that no amount of additional axioms can resolve the problem. He established, in other words, that mathematics is *inherently incomplete*, and that certain truths will remain unknowable to us. This was (and continues to be) a very surprising conclusion, since one intuitively expects that it is always possible to establish whether or not a mathematical proposition is true (at least, in principle).

The fact that certain propositions defy this expectation has a number of important consequences. Among other things, it suggests that "automatic" proof verification is impossible, and that formal algorithms can never completely replace qualified experts when it comes to validating theorems. Under such circumstances, there are no guarantees that a consensus will always be achieved, or that subtle errors will not be uncovered at some point in the future. We must concede, in other words, that mathematics cannot be separated from the human thought process, and that this discipline is not infallible.

This inherent "imperfection" of mathematics raises an interesting practical question. In the absence of absolute certainty, what criteria must be met before a proposition can be accepted as *true*? Will mathematics remain "pure", and recognize only statements that have been rigorously proved, or will its practitioners eventually adopt the model used in the natural sciences, where hypotheses are accepted as valid when they are shown to be sufficiently probable? The fact that suggestions of this sort (which would have thoroughly shocked mathematicians of centuries past) are now being seriously discussed suggests that we are witnessing a significant change in the way mathematics is practiced.

Another development that has affected the way we think about mathematics is the emergence of various "experimental" techniques, which are based on computer-aided simulation. This method of investigation has gained wide acceptance in recent years, due in large part to its ability to uncover new and unexpected problems. The potential value of this approach is particularly evident in areas such as chaos theory and fractal geometry, neither of which existed (or could have existed) prior to the advent of computers. It goes without saying, of course, that mathematical results anticipated through simulation still require a rigorous proof before they can be accepted as more than conjectures. In that respect, mathematics is still the same as it was centuries ago. It is also true, however, that experimental techniques open up a range of previously unanticipated possibilities, and can often provide us with a very accurate idea of what it is that we should attempt to prove. If this methodology continues to attract new adherents (which seems very likely), it will become increasingly inappropriate to view mathematics as a purely abstract and non-empirical discipline.

In thinking about the impact of Gödel's theorem, it is also important to emphasize that this is by no means the only such result in mathematics. Turing's Halting Problem in computer science, the Continuum Hypothesis in set theory and Chaitin's constant Omega all point to the existence of unprovable propositions in mathematics. The case of Omega (which we will consider in greater detail in Section 2.4) is particularly interesting in this context, since it has been rigorously shown that we can determine only a very limited number of its bits. The remaining bits in its binary expansion elude computation, and no new theory or improvement in computing power will ever allow us to deduce them. Each of these bits can therefore be viewed as an "unknowable truth", much like the one discovered by Gödel.

In the following sections we will take a closer look at some of the issues that were raised in this brief introduction. We begin with a description of certain important properties of formal systems, such as internal consistency and completeness. This will provide the necessary background for understanding Gödel's theorem, whose proof is outlined in Section 2.3. We will then examine how Gödel's work affected other areas of modern mathematics, such as set theory, non-standard analysis and algorithmic information theory.

2.1 An Introduction to Formal Systems

A formal system can be fully defined by a set of axioms and two types of rules:

RULES OF FORMATION, which determine whether or not a string of symbols is a legitimate member of the system.

RULES OF INFERENCE, which provide the basis for generating new theorems from axioms and already existing theorems.

To illustrate the main properties of such systems, let us once again consider the so-called PQ system[1], which consists of three types of symbols - letters P, Q, and strings of hyphens. The rules and axioms of this system can be described in the following manner.

Rule of Formation. Any string of the type

$$x\,P\,y\,Q\,z \tag{2.1}$$

where x, y and z represent sequences of one or more hyphens is a member of this system (we will refer to it as a *well-formed string*).

Axioms. Every string of the type

$$x\,P - Q\,x- \tag{2.2}$$

where x is a sequence of hyphens is an axiom.

Rule of Inference. If

$$x\,P\,y\,Q\,z \tag{2.3}$$

is a theorem, then

$$x\,P\,y - Q\,z- \tag{2.4}$$

is also a theorem.

Two remarks are in order before we proceed.

Remark 2.1. The rule of formation implies that not every combination of P, Q and hyphens is a member of this system. For example,

$$- - -P - -Q - -- \tag{2.5}$$

is a well-formed string, but

$$- - -P\,Q - -- \tag{2.6}$$

or

$$-P - Q - P \tag{2.7}$$

are *not*.

Remark 2.2. The PQ system has infinitely many axioms, since x can be *any* sequence of hyphens.

The theorems of the PQ system can be automatically generated from the axioms, by applying the rule of inference. For instance, starting with axiom

$$- - P - Q - -- \tag{2.8}$$

we can generate the following theorem

$$- - P - -Q - - - - \tag{2.9}$$

Applying the rule of inference again, we get

$$- - P - - - Q - - - - -- \tag{2.10}$$

and so on. It would be rather straightforward to program a computer to generate theorems in this manner. Note, however, that such a program would obviously never terminate, since there are infinitely many axioms and boundless possibilities to produce theorems from them.

In the following, we will use this simple example to introduce two fundamental properties that characterize any formal system - *internal consistency* and *completeness.* These properties will now be described in greater detail, as will some of their most important implications. Such an analysis will provide the necessary background for understanding Gödel's Incompleteness Theorem and its consequences.

Internal Consistency

A key problem facing any formal system is the verification of *internal consistency*. The main issue that arises in this context is summarized by the following question: How can one be sure that the rules of inference don't produce two theorems that contradict each other? Answering such a question can be a formidable task, even in simple cases. The problem is compounded by the fact that consistency can be defined in a number of different ways (which are equivalent for most reasonable systems, but are not identical in general).

We begin with a definition that assumes that the system has some form of *negation* built into its structure.

Definition 2.1. A formal system is said to be internally consistent if there is no well formed string P such that both P and its negation, $\sim P$, are theorems.

Proofs of consistency in the sense of Definition 2.1 can be very difficult. If we attempted to do this *within* the formal system, our only option would be to generate theorems until we find two that stand in logical opposition. In this way, we might be able to establish *inconsistency* in a finite number of steps (if we are very, very fortunate). Consistency, however, can never be verified in this manner, since infinitely many theorems need to be examined. It follows, therefore, that a proof of consistency requires reasoning that *transcends* the formal system.

One possible approach to this problem would be to consider an alternative definition of consistency, which is based on the notion of an *interpretation*. An interpretation can be viewed as a mapping that "translates" all the statements about a particular formal system S into statements about some other (preferably more manageable) system M.

Definition 2.2. A formal system is said to be internally consistent if there exists an interpretation in which every one of its theorems translates into a true statement.

In order to frame Definition 2.2 in slightly more technical terms, let Φ denote the mapping from S into a model system M, and assume that $\Phi(S) = W \subset M$. This is equivalent to saying that all the statements of S map into a *subset* of statements about M. If we now denote all the theorems contained in S and W by T_S and T_M, respectively, then consistency requires

$$\Phi(T_S) \subset T_M \tag{2.11}$$

Condition (2.11) is illustrated in Fig. 2.1, which schematically shows how *every* theorem of S must "translate" into a theorem of M in order for S to be a consistent system.[2]

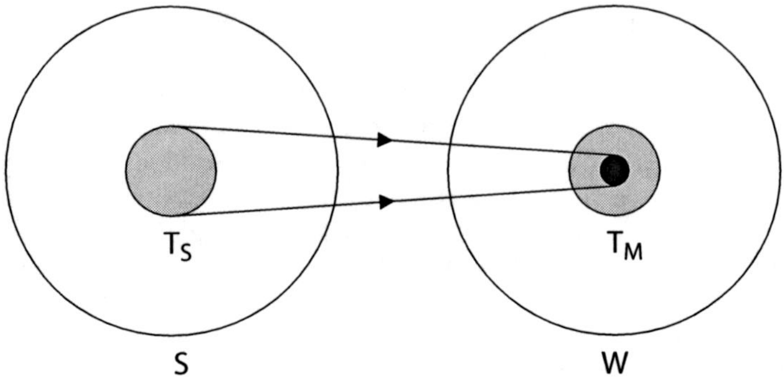

Fig. 2.1. A schematic illustration of consistency.

To see the practical value of Definition 2.2, let us assume that P, Q and sequences of hyphens are interpreted in the following way:

Symbol	Interpretation	
P	$+$	
Q	$=$	
$-$	1	(2.12)
$--$	2	
$---$	3	
$\vdots$	$\vdots$	

According to (2.12), every one of our axioms can now be translated into a statement of the form

$$N + 1 = N + 1 \tag{2.13}$$

which is clearly a truth about integers. Furthermore, under this interpretation the rule of inference preserves the truth of the statement. To see this more clearly, recall that if $x\,PyQ\,z$ is a theorem, then $x\,P\,y-Q\,z-$ must be a theorem as well. This property obviously extends to the translations of these statements, since the truth of $x + y = z$ implies the truth of $x + y + 1 = z + 1$. Since all theorems are derived using this rule of inference, we can claim that the PQ system is internally consistent.

Although the consistency proof proposed above is very simple, it is important to recognize that its logical validity *does not* follow exclusively from the properties of the PQ system. Instead, it requires a certain amount of *external* reasoning as well, which relies on verifiable facts from elementary number theory. This is a pattern that we will encounter in all subsequent applications of Definition 2.2.

Completeness

A related (and equally important) problem associated with a formal system amounts to establishing whether or not it is *complete*. As in the case of consistency, here too there are a number of possible definitions, two of which we provide below.

Definition 2.3. A formal system is said to be complete if for any well formed string P, either P or its negation, $\sim P$, is a theorem.

According to this definition, a system is *incomplete* if neither P nor $\sim P$ are theorems. If such a proposition P exists, it is said to be *undecidable* (this is a notion that Gödel used extensively in establishing his results).

An alternative definition of completeness is a *relative* one, and is provided with respect to a given interpretation.

Definition 2.4. A formal system S is complete with respect to a given interpretation if *every* truth that is expressible in S is also a theorem.

In terms of our previous notation, the completeness of S with respect to a model system M entails

$$\Phi(T_S) = T_M \tag{2.14}$$

which is a stronger condition than (2.11). What (2.14) requires is that *every* theorem (*i.e.* true statement) in M that can be translated into a statement in S corresponds to a theorem in S. To better understand what this means, consider the scenario where there exists a theorem t_0 in T_M that maps into some $p \notin T_S$. In that case, system S would be *incomplete* with respect to this interpretation (a schematic illustration of such a situation is provided in Fig. 2.2).[3]

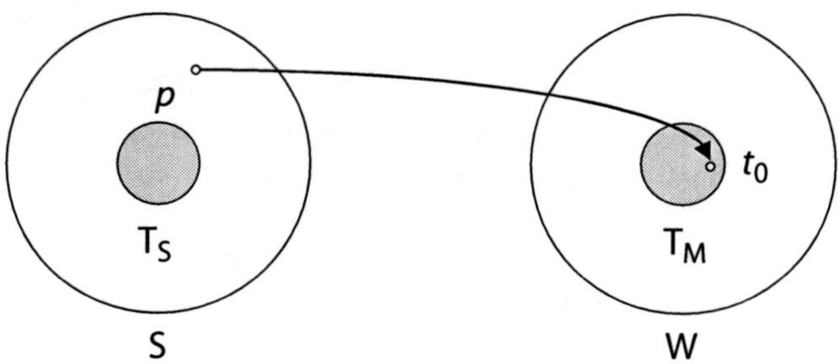

Fig. 2.2. A schematic illustration of incompleteness.

In the case of the PQ system and the interpretation proposed in (2.12), the problem of completeness reduces to the following question: Is every expressible truth about adding two integers also a theorem? It turns out that this is indeed the case. To see that, consider three arbitrary integers a, b and c that satisfy

$$a + b = c \tag{2.15}$$

(this is the only type of algebraic formula that can be represented in the PQ system using our interpretation). If we start with axiom

$$x\,P - Q\,x- \tag{2.16}$$

where x is a string of a hyphens and apply the rules of inference $b-1$ times, we will obtain

$$x\,P\,y\,Q\,z \tag{2.17}$$

Theorem (2.17) obviously corresponds to expression (2.15), since x, y and z are strings consisting of a, b and $c = a + b$ hyphens, respectively. The proposed interpretation of the symbols thus allows us to prove both the completeness and the internal consistency of the PQ system.

The Modified PQ System

The fact that the consistency and completeness of a formal system (in the sense of Definitions 2.2 and 2.4) can be established using an interpretation of the symbols suggests that these properties needn't be absolute. To get an idea of some of the issues that arise in this context, let us consider a modified version of the PQ system, in which we introduce a second set of axioms:

AXIOM SCHEME 2. Every string of the type

$$x\,P-Q\,x \tag{2.18}$$

is also an axiom.

Under the old interpretation of the symbols, the axiom

$$-P-Q- \tag{2.19}$$

would be interpreted as

$$1+1=1 \tag{2.20}$$

which is clearly false. Does this mean that the modified system is actually inconsistent? Not necessarily, if we can suitably *reinterpret* the symbols. To that effect, let use consider the following translation:

Symbol	Interpretation
P	$+$
Q	$\geq$
$-$	1
$--$	2
$---$	3
$\vdots$	$\vdots$

(2.21)

The general statement

$$x\,P-Q\,x \tag{2.22}$$

would now correspond to

$$N+1 \geq N \tag{2.23}$$

which is obviously correct. Since Definition 2.2 requires that $\Phi(T_S) \subset T_M$ in *at least one* interpretation, it follows that the PQ system remains consistent after adding Axiom Scheme 2.

What about completeness? The true arithmetical statement

$$2 + 3 \geq 1 \tag{2.24}$$

can obviously be expressed as

$$- - P - - - Q- \tag{2.25}$$

which is a well-formed string, but is *not* a theorem in the PQ system. Indeed, if we wanted to 'backtrack' toward an axiom of the form

$$x\, P - Q\, x \tag{2.26}$$

or

$$x\, P - Q\, x- \tag{2.27}$$

we would need to reduce the number of hyphens between P and Q to one. The rule of inference requires that we reduce the number of hyphens on *both* sides of Q by 1 in each step, so after one step (2.25) would become

$$- - P - -Q \tag{2.28}$$

Note, however, that this is not a well formed string any more.

The above discussion suggests that the modified PQ system is consistent, but is *incomplete* with respect to the interpretation proposed in (2.21). This follows directly from the fact that there are truths about integers that can be expressed as well formed strings, but are not theorems. However, it is entirely possible that we could come up with a different interpretation of the symbols which would guarantee *both* consistency and completeness. For the modified PQ system, it turns out that there is such an interpretation. The only necessary modification would be to reinterpret Q as 'equal or greater by one' (you are welcome to try this!).

2.2 Absolute Proofs

Establishing the consistency and completeness of complicated formal systems can be an extraordinarily difficult task. Even if one could find an interpretation that works, proving completeness requires showing that *all* expressible truths are also theorems. In most cases, this would entail an infinite number of demonstrations, which is obviously impossible in practice (remember that although any finite number of tests may be positive, this doesn't allow us to conclude that there won't be unpleasant surprises down the line). In this section, we will consider a rather elaborate formal system for which it is actually possible to show both consistency and completeness without interpreting all the symbols. This system (often referred to as Sentential Calculus) is interesting because it represents a subset of the formal system that inspired Gödel's theorems. It will also illustrate the flavor (if not the details) of Gödel's proof, and show that establishing properties such as consistency requires *metamathematical* techniques (that is, reasoning 'outside' the formal system).

Sentential Calculus

Sentential Calculus consists of two distinct types of symbols, which are referred to as variables and connectives. A brief description of each type is provided below.[4]

Variables are simply letters of the alphabet, with no interpretation associated with them.

Connectives are additional symbols that have a specific interpretation, and are used to link variables (and/or strings). They are defined as follows:

Symbol	Interpretation
$\sim$	not
$\vee$	or
$\supset$	if-then
$\wedge$	and

(2.29)

Rules of Formation. There are three rules of formation:

(*i*) Each letter of the alphabet is a well formed string.

(*ii*) If S is a well-formed string, then so is $\sim S$.

(*iii*) If S_1 and S_2 are well formed strings, then so are:

$$\begin{gathered} S_1 \vee S_2 \\ S_1 \supset S_2 \\ S_1 \wedge S_2 \end{gathered} \tag{2.30}$$

Based on these rules, it is easy to verify whether a string is well-formed or not. For example, strings

$$\begin{gathered} p \\ \sim p \\ p \vee q \\ (p \vee q) \supset (q \vee p) \end{gathered} \tag{2.31}$$

are all well formed, while

$$\begin{gathered} p \sim q \\ q\,p\vee \\ p\wedge \end{gathered} \tag{2.32}$$

are not.

Axioms. Sentential calculus has four axioms:

$$\begin{array}{ll} \text{Axiom 1} & (p \vee p) \supset p \\ \text{Axiom 2} & p \supset (p \vee q) \\ \text{Axiom 3} & (p \vee q) \supset (q \vee p) \\ \text{Axiom 4} & (p \supset q) \supset [(r \vee p) \supset (r \vee q)] \end{array} \tag{2.33}$$

Note that since the variables have no interpretation, these axioms have no specific meaning.

Rules of Inference. The rules of inference determine how new theorems are obtained from existing ones. In this system, there are two such rules: the Rule of Substitution and the Rule of Detachment.

(*i*) **Rule of Substitution**. Suppose that string S_1 is a theorem, and that we substitute *every* occurrence of a certain variable with some well-formed string S_2. Then, the resulting string is itself a theorem. As an illustration, consider the string

$$(p \vee q) \supset (q \vee p) \tag{2.34}$$

(which is actually an axiom). If we were to replace p with some arbitrary well-formed string S, the resulting statement

$$(S \vee q) \supset (q \vee S) \tag{2.35}$$

would be a theorem (although S itself needn't be).

(*ii*) **Rule of Detachment**. Suppose that S_1 and $S_1 \supset S_2$ are both theorems. Then, S_2 is a theorem as well.

Now that we have introduced all the ingredients of the formal system, our objective will be to prove that it is internally consistent. The proof involves a sequence of steps, and *does not* require an interpretation of the variables.

STEP 1. As a first step, let us consider what would happen if two strings S_1 and $\sim S_1$ *both* turned out to be theorems. In that case, a statement and its negation would be provable simultaneously, which clearly implies inconsistency (in the sense of Definition 2.1). We will now show that if Sentential Calculus were indeed inconsistent in this way, then *any* well-formed string would necessarily be a theorem. In order to do this, we will assume that both S_1 and $\sim S_1$ are theorems, and make use of the fact that

$$p \supset (\sim p \supset q) \tag{2.36}$$

is a theorem in our system (we will not prove this). Applying the Rule of Substitution twice, we obtain

$$S_1 \supset (\sim S_1 \supset S) \tag{2.37}$$

which is a theorem for *any* well-formed string S. The rule of Detachment further implies that

$$\sim S_1 \supset S \tag{2.38}$$

is a theorem as well. We now have a rather interesting situation, in which both $\sim S_1$ and $\sim S_1 \supset S$ are theorems. Applying the Rule of Detachment to these two statements, we conclude that S is itself a theorem, although it was an arbitrarily chosen string. In other words, if Sentential Calculus were an inconsistent system, then *any* well-formed string in it would be a theorem.

STEP 2. In view of the result obtained in Step 1, we can now claim that Sentential Calculus is consistent if there exists *at least one* well-formed string that is not a theorem. The trick, of course, is to find such a statement, and this is by no means easy. A necessary step in this process is to identify a property that is common to all axioms and theorems in the system. In order to do that, let us group all the letters of the alphabet into two classes, denoted K_1 and K_2. The compound strings will then be classified according to the following rules:

Rule 1. A string of the form $S_1 \vee S_2$ is placed in class K_2 if *both* S_1 and S_2 are in K_2. Otherwise, it is placed in K_1.

Rule 2. A string of the form $S_1 \supset S_2$ is placed in class K_2 if S_1 is in K_1 and S_2 is in K_2. Otherwise, it is placed in K_1.

Rule 3. A string of the form $S_1 \wedge S_2$ is placed in class K_1 if *both* S_1 and S_2 are in K_1. Otherwise, it is placed in K_2.

Rule 4. A string of the form $\sim S$ is placed in class K_2 if S is in K_1. Otherwise, it is placed in K_1.

Based on this classification, we will say that a string S is a *tautology* if it is a member of class K_1 regardless of where its constitutive elements belong.

STEP 3. In this step we will show that all the axioms of Sentential Calculus are tautologies. We will demonstrate this property explicitly for Axioms 1 and 2 (the proof is similar for Axioms 3 and 4). The easiest way to do this is to tabulate all the possible combinations. For Axiom 1, we have

p	$(p \vee p)$	$(p \vee p) \supset p$
K_1	K_1	K_1
K_2	K_2	K_1

(2.39)

while Axiom 2 corresponds to

p	q	$(p \vee q)$	$p \supset (p \vee q)$
K_1	K_1	K_1	K_1
K_1	K_2	K_1	K_1
K_2	K_1	K_1	K_1
K_2	K_2	K_2	K_1

(2.40)

It is obvious that both axioms always belong to class K_1, regardless of how we classify letters p and q.

STEP 4. Having established that all axioms are tautologies, we now proceed to show that this is true of all theorems as well. Since theorems are derived from axioms by the two rules of inference, it suffices to prove that these rules preserve the 'tautology' property. In the following, we will demonstrate this only for the Rule of Detachment (the proof for Rule of Substitution is similar).

Suppose that S_1 and $S_1 \supset S_2$ are *both* theorems *and* tautologies. This means that $S_1 \in K_1$ and $S_1 \supset S_2 \in K_1$ no matter where their constitutive elements belong. The Rule of Detachment implies that S_2 must be a theorem as well, but it is unclear at this point whether or not it is a tautology.

What follows is an example of a proof by contradiction. Namely, let us assume that S_2 *is not* a tautology. In that case, there exists a possible combination of constitutive elements for which $S_2 \in K_2$. For that particular combination, we have $S_1 \in K_1$ and $S_2 \in K_2$, and Rule 2 implies that $S_1 \supset S_2 \in K_2$ as well. This is a contradiction, since we assumed that $S_1 \supset S_2$ is a tautology. As a result, our original assumption must be wrong, and S_2 must be a tautology whenever S_1 and $S_1 \supset S_2$ have this property.

STEP 5. Since all axioms and theorems were shown to be tautologies, our task now reduces to finding a well-formed string that is *not*. This is actually quite simple. An obvious example would be the string

$$p \vee q \tag{2.41}$$

which is clearly well-formed, but is not a tautology. Why this is so is clear from the table below, in which all possible combinations are examined.

p	q	$(p \vee q)$
K_1	K_1	K_1
K_1	K_2	K_1
K_2	K_1	K_1
K_2	K_2	K_2

(2.42)

Conclusions

The results of this section suggest that consistency can sometimes be established *without* resorting to specific interpretations. When such a proof can be found, it is referred to as *absolute*. It is important to note in this context that our methodology once again relied on reasoning 'outside' the formal system.[5] The only tools available within the Sentential Calculus are the two Rules of Inference, and these were clearly not sufficient to establish consistency. Proofs of this type are called *metaproofs*, since they transcend the limits set by the formal system.

Remark 2.3. An absolute proof of completeness is also possible for Sentential Calculus, but we will not consider it here.

2.3 Gödel's Incompleteness Theorem

Based on the discussion in Sections 2.1 and 2.2, we can now consider Gödel's famous Incompleteness Theorem and provide an outline of its proof. Given the complexity of this proof, in the following we will focus only on its main ideas[6].

We begin with a description of Gödel's main result, which can be summarized in the following way.

Theorem 2.1. (The First Incompleteness Theorem) In every sufficiently complex formal system there are unprovable propositions. Even if such a system were augmented by an indefinite number of new axioms and rules, there would always be number theoretic truths that are not formally derivable.

Theorem 2.1 has two important corollaries, which pertain to consistency and completeness.

Corollary 2.1. It is impossible to prove the consistency of the formal system proposed in Principia Mathematica, or any similarly complex formal system within the system itself.

Corollary 2.2. Given any consistent formalization of number theory (including the one in Principia Mathematica), there are true number theoretic statements (whose truth can be established by external means) that are expressible as well-formed strings, but are *not* theorems. In other words, no sufficiently complex formal system is complete with respect to the "standard" interpretation (which assumes number theory as its model).

To get a sense for how the Incompleteness Theorem was proved, we first need to briefly describe a formal system known as Typographical Number Theory (TNT), which was designed to represent statements about non-negative integers.[7] This system uses a limited number of standard arithmetical and logic symbols, which are listed in Table 2.1. Note that all positive integers can be described by combining prefix S and number zero in the manner indicated in Table 2.2.

As noted earlier, one of the key elements of Gödel's proof is the construction of a mapping that associates a specific integer with every TNT symbol. A possible way to do this is shown in Table 2.1, where each symbol is matched to a triplet of digits from the set $\{1,2,3,6\}$.[8]

Using this mapping, we can *uniquely* express any TNT string as an integer (usually, a very large one). For example, the string

$$\mathrm{S0} + 0 = \mathrm{S0} \tag{2.43}$$

corresponds to the 21 digit number $\alpha = 123,666,112,666,111,123,666$, which will be referred to in the following as the *Gödel number* of (2.43).[9]

Before proceeding, it is important to add that TNT statements can include one or more *variables* (which are denoted a, a', a'', etc.). Such statements fall into two general categories, which are known as *open* and *closed* formulas. An example of an open formula would be the string

$$\mathrm{S0} + a = \mathrm{S0} \tag{2.44}$$

where a is a *free variable.* In cases like this, we cannot determine whether the statement is true or false, since a is unspecified.

Closed formulas are characterized by the fact that each variable has an associated *quantifier.* This requirement ensures that the truth of the statement can always be decided. As an illustration of this property, consider a modification of (2.44) where the quantifier $\exists a$: is added at the beginning of the string. It is easily verified that the resulting formula

$$\exists a : \mathrm{S0} + a = \mathrm{S0} \tag{2.45}$$

represents a true statement, whose interpretation is: "There exists a non-negative number a such that $1 + a = a$."

The Gödel numbering defined in Table 2.1 allows us to represent an entire proof as a single integer. Indeed, since formal proofs proceed as sequences of individual statements, all we need to do is find the Gödel number for each of them, and separate consecutive lines by inserting the triplet 611. Note that the last line of such a derivation always corresponds to the actual theorem that we set out to prove. This information is easily extracted from the Gödel number of the proof by identifying all digits that follow the last appearance of 611. For example, if a proof were represented by Gödel number

$$\beta = 626,262,636,\ldots,611,333,262,636,123,666,112,262,111,123,666 \tag{2.46}$$

the theorem itself would have the Gödel number

$$\gamma = 333,262,636,123,666,112,262,111,123,666 \tag{2.47}$$

A number like γ which corresponds to a TNT theorem will be referred to in the following as a *TNT number.*

In order to provide a compact description of Gödel's proof, we need to introduce two special number theoretic functions, both of which are can be represented as TNT strings. The first one, denoted $F_1(a, a')$, takes a (which is the Gödel number of an entire proof) and checks if a' is the Gödel number of its

Number	TNT Symbol
0	0
1	S0
2	SS0
3	SSS0
⋮	⋮

Table 2.1. Representation of numbers in TNT.

TNT Symbol	Integer Representation
0	666
S	123
$=$	111
$+$	112
$\cdot$	236
(	362
)	323
$<$	212
$>$	213
]	312
[	313
a	262
$'$	163
$\wedge$	161
$\vee$	616
$\supset$	633
$\sim$	223
$\exists$	333
$\forall$	626
:	636
end of line	611

Table 2.2. Integer representation of TNT symbols.

last line. In view of our previous discussion, it follows that a' is a TNT number whenever $F_1(a, a')$ is true for some a.

The second function, denoted $F_2(a'', a')$, is slightly more complicated. Its first argument, a'', is assumed to be the Gödel number of a known TNT string with a single free variable. This function replaces every occurrence of the free variable with a'', and then checks whether a' is the Gödel number of the resulting string.

Given these two definitions, let us now consider the string H:

$$\sim \exists a : \exists a' : \langle S_1(a, a') \wedge S_2(a'', a') \rangle \tag{2.48}$$

where $S_1(a, a')$ and $S_2(a'', a')$ are TNT representations of functions $F_1(a, a')$ and $F_2(a'', a')$, respectively. Note that a'' is a free variable in H, since it has no quantifier associated with it. Assuming that the Gödel number of (2.48) is u, we will now form a new string G by replacing every occurrence of a'' in H by u:

$$\sim \exists a : \exists a' : \langle S_1(a, a') \wedge S_2(u, a') \rangle \tag{2.49}$$

The Gödel number for string G will be denoted in the following by g (which implies that $F_2(u, g)$ is true by definition).

The number theoretic interpretation of G is: "There do *not* exist numbers a and a' such that $F_1(a, a')$ and $F_2(u, a')$ are simultaneously true." This interpretation gives rise to two possible scenarios.

CASE 1. Suppose that the number theoretic interpretation of G is a *false* statement. Then, there exist a and a' such that *both* $F_1(a, a')$ and $F_2(u, a')$ are true. By definition, $F_2(u, a')$ is true only when $a' = g$. Therefore, $F_1(a, g)$ must be true for some a. This means that g is a TNT number, which is equivalent to saying that G is a TNT theorem. Such a result represents a contradiction, since the interpretation of a TNT theorem cannot be a false statement (bear in mind that every TNT theorem must map into a number theoretic truth).

CASE 2. Suppose that the number theoretic interpretation of G is a *true* statement. This means that there is *no* pair of numbers a and a' such that $F_1(a, a')$ and $F_2(u, a')$ are simultaneously true. Let us now consider the special case when $a' = g$. Since $F_2(u, g)$ is true by definition, then $F_1(a, g)$ must be false for *all* a. This means that g is *not* the last line of a TNT proof, and therefore cannot be a TNT number. In other words, G is *not* a TNT theorem.

There are no contradictions in Case 2, but the implication is that G corresponds to a number theoretic truth although it is *not* a TNT theorem. This represents incompleteness as we defined it in Section 2.1. It is interesting to observe that the interpretation of G actually tells us something about the string itself (specifically, that G is not a TNT theorem). This suggests that the incompleteness of TNT has its roots in the possibility of *self-reference*.

2.4 Incompleteness in Modern Mathematics

In this section we will consider several areas of modern mathematics whose development was greatly influenced by Gödel's work. Some of the results obtained in these fields challenge our basic intuitive concepts, and extend them in ways that open up exciting new possibilities. We will begin with an example from computer science, and then move on to a discussion of non-Cantorian set theory and non-standard analysis.[10]

The Elusive Number Omega

In many ways, a formal system is very much like a computer program. Indeed, one can easily envision the axioms and rules of inference as an *algorithm*, whose output are the resulting theorems. With that in mind, it is not very surprising that a form of Gödelian incompleteness is encountered in the domain of computer science. In this section, we will concentrate on one such case, which involves the number Omega introduced by Gregory Chaitin.[11]

In order to understand what makes Omega so interesting, we first need to establish some basic facts about computer programs. We will begin by considering the following simple code, whose purpose is to compute the smallest number n for which the sum

$$s_n = \sum_{k=0}^{n} \frac{1}{2^k} \tag{2.50}$$

exceeds some prescribed value m.

Algorithm 2.1. The sum described in (2.50) can be encoded as a Matlab m-file in the following way:

```
function [n, s] = program1(m)
s = 0; k = 0;
while(s < m)
      s = s + 2^(-k);
      k = k + 1;
end
n = k;
```

As the program repeatedly executes this loop, the variable s takes on values 1, 1.5, 1.75, 1.875 etc. Consequently, if we set $m = 1.6$ in this program, it will halt after passing through the loop three times, and will return $n = 3$ and $s = 1.75$ as its output. If, on the other hand, we were to set $m = 2.6$, the program would *never halt.* This follows directly from the fact that s_n increases monotonically, with 2 as its upper limit.

The distinction between programs that halt and those that do not is at the heart of one of the most profound results in computer science. This result relates to the so-called Halting Problem, which was posed (and solved) by Alan Turing in the 1930s, when computers were no more than a thought experiment. Turing's question can be paraphrased as follows: "Is there a program that can decide in a finite amount of time whether any other program will halt?" In showing that such a program does *not* exist, Turing discovered a kind of incompleteness akin to the one established by Gödel. Indeed, one of the immediate consequences of this result is that there will always be a program for which we cannot formally establish whether or not it will halt.

To get a better sense for why the Halting Problem cannot be resolved by a computer program, it is useful to consider a simple formal system of the kind introduced in Section 2.1. Such a system consists of axioms and rules of inference, which produce strings of characters as theorems. Let us now assume that we are given a particular string, and are asked whether or not it is a theorem. How could we tell? If we attempt to find an answer within the formal system itself, the only way to proceed would be to generate the theorems one by one, and hope that one of them will ultimately match the given string. However, if this does not happen, the process could go on forever. In other words, such a formal system cannot establish the "theoremhood" of any given string in a finite amount of time. A similar (although much more subtle) kind of reasoning lies behind Turing's result as well.

The number Omega is intimately related to the Halting Problem. In order to define it, let us consider how one could *randomly* generate a program. A simple way to do this would be to flip a coin whenever the computer requests the next bit of the program. This bit will be a 1 if the outcome of the toss is "heads," and 0 if it is "tails." The process continues until the program halts, at which point the computer stops asking for additional bits. Programs with this property are called *self-delimiting*, due to the fact that they implicitly control their own length.

Let us now denote the set of all self-delimiting programs that halt by H, and form the sum

$$\Omega = \sum_{p \in H} 2^{-|p|} \tag{2.51}$$

where $|p|$ represents the length of program p measured in bits. The constant Ω defined in this way corresponds to the probability that a randomly chosen program will halt. To justify this interpretation, suppose that among all possible self-delimiting programs only 111 and 0101 actually halt. If we adopt the coin flipping model for generating random programs, then the halting probability is the sum of the probability that the first three tosses produce 111 (which is 2^{-3}) and the probability that the first four tosses produce 0101 (which is 2^{-4}). Note that this is precisely the way in which Ω is constructed.[12]

To better understand some of the key properties of Ω, we need to introduce the notion of *algorithmic complexity*, which can be traced back to the 17th century and the great German mathematician and philosopher Gottfried Wilhelm von Leibniz. In his *Discourse on Metaphysics*, Leibniz pondered how one could differentiate observations that conform to a law from those that are random. To gain some insight into this question, consider a set of n randomly selected points in a plane. It is well known that one can always construct an interpolation polynomial of order n or higher that passes through all these points. The difficulty, however, is that such a polynomial is no simpler than the data that it is trying to "explain." As a result, this mathematical function tells us nothing new or useful. In view of this problem, Leibniz proposed that we should use *simplicity* as a criterion for distinguishing between phenomena that are lawful and those that are not.

In the context of algorithmic information theory, the criterion proposed by Leibniz can be reformulated in terms of *program length*. To see how that works, suppose that we are given the first N decimal digits of a real number, and are asked to find the *smallest* self-delimiting program that will produce this data as its output. If we succeed, the size of this program (measured in bits) corresponds to the *complexity* of the output. The following example is designed to illustrate this idea more clearly.

Example 2.1. Let us consider the first N digits in the decimal expansion of π (where N is a large, fixed number). How complicated does the program have to be in order to produce such a output? It turns out that the required program is actually quite simple. One possibility is based on the following formula established by David and Gregory Chudnovsky:

$$\frac{1}{\pi} = 12 \sum_{k=0}^{\infty} \frac{(-1)^k (6k!)(13,591,409 + 545,140,134 \cdot k)}{(3k!)(k!)^3 640,320^{3k+1.5}} \tag{2.52}$$

The formula looks rather messy, but it can be coded in just a few lines.

It can be shown that each term in the sum (2.52) produces an additional 14 digits of π. More importantly, the length of the code is completely independent of N (the execution time is a different matter, but that has nothing to do with our definition of complexity). With that in mind, we can conclude that the enormous amount of raw data contained in the digits of π carries little in the way of algorithmic information. All this information can actually be *compressed* into a very simple code, since the decimals of π obey a certain pattern that is captured by (2.52).[13]

In the case of Ω, the situation is very different from the one in Example 2.1. It turns out, for instance, that in order to compute the first N bits in its binary expansion, we need a program whose size is close to N bits. This means that the algorithmic information contained in the bits of Ω *cannot* be compressed into a program that is substantially smaller than the data itself. Things are actually even worse than that. For real numbers such as π (which are described by explicit formulas) we can increase the number of computed digits indefinitely. We cannot do this for Ω, although it is a well defined number with a definite binary expansion. As Chaitin pointed out, each of the unknowable bits of Ω represents a concrete example of Gödel's incompleteness.

> "Gödel's incompleteness theorem tells us that within mathematics there are statements that are unknowable, or undecidable. Omega tells us that there are in fact infinitely many such statements: whether any one of the infinitely many bits of Omega is a 0 or a 1 is something that we cannot deduce from any mathematical theory. More precisely, any maths theory enables us to determine at most finitely many bits of Omega." *Gregory Chaitin*[14]

What makes Ω particularly interesting is the fact that it is *not* an abstract creation like Gödel's unprovable proposition. Omega is an actual number, about which we know quite a bit and can precisely specify how it can be constructed. It seems that we are tantalizingly close to computing it, and yet we know that it will remain eternally elusive.

> "I wouldn't say that we can touch it, but we can certainly point straight at it.... That's much more frustrating and much more interesting than being told that we can't have something or do something that never seemed concrete enough or down to earth." *Gregory Chaitin*[15]

Non-Cantorian Set Theory

The concept of a set is one of the oldest and most basic in all of mathematics. For a very long time, this notion was accepted intuitively. Cantor's definition proposed in the 19th century is a typical illustration of this approach:

Definition 2.5. A set is any collection into a whole of definite and separate objects of our intuition or thought.

It turns out, however, that Cantor's idea of a set leaves some fundamental questions unanswered. In the following, we will focus on two propositions that have been particularly troublesome in that respect - the Axiom of Choice and the Continuum Hypothesis.

To properly understand what these propositions claim, we must first recognize that there are different kinds of infinite sets, and that the sizes of these sets can be compared in a meaningful way. It can be shown, for example, that the set of all real numbers is in some sense "larger" than the set of all integers. In technical terms, the two sets are said to have *cardinality* c and $\aleph_0$ (pronounced aleph zero), respectively, with $c > \aleph_0$. Another way to make this distinction is to say that the set of integers is *countable*, while the set of real numbers is *not*.

We now turn to the Axiom of Choice, which can be formulated as follows:

P1: *The Axiom of Choice.* Let α be a collection of sets $\{A, B, \ldots\}$, and assume that none of these sets is empty. The axiom of choice states that there exists a set Z consisting of precisely one element each from A, B, etc.

This proposition has been viewed as problematic by many mathematicians, much like Euclid's 5th axiom. To see why this is so, let us consider the simple set $A = \{1, 2, 3\}$, and form a set $B = \{\{\}, \{1\}, \{2\}, \{3\}, \{1, 2\}, \{1, 3\}, \{2, 3\}, \{1, 2, 3\}\}$ which consists of all its subsets. Set B is referred to as the *power set* of A, and is usually denoted by $B = 2^A$. This notation reflects the fact that any finite set A with m elements has exactly 2^m subsets (a property that is easily verified for our sets A and B). It is obvious that proposition P1 holds true in all such cases, but it remains unclear whether this conclusion can be extended to infinite sets as well. In order to appreciate the complexity of the question, it suffices to observe that the set of integers (whose cardinality is $\aleph_0$) contains *uncountably many* subsets. This property was proved by Georg Cantor, and is formally expressed as $c = 2^{\aleph_0}$. Under such circumstances, it is difficult to see how proposition P1 could be verified in a systematic manner.

The second proposition that we will be concerned with addresses the question of whether there are infinite sets whose cardinality lies *between* $\aleph_0$ and c. Cantor assumed that such sets do *not* exist, and this conjecture is referred to as the "continuum hypothesis". Determining whether or not this hypothesis is correct proved to be a very difficult task – so difficult, in fact, that it became one of the most celebrated unsolved problems in mathematics (it was the first one on Hilbert's famous list, which was compiled around 1900).

P2: *The Continuum Hypothesis.* In an infinite set, all infinite subsets have either cardinality $\aleph_0$ or c.

As a first step in addressing these difficult issues, it is necessary to recast the notion of a set in more formal terms. The appropriate axiomatic framework was introduced in the 1920s by the Norwegian logician Thoralf Skolem, based on earlier work by Ernst Zermelo and Abraham Fraenkel. This system (without P1 and P2) is commonly referred to as Restricted Set Theory, which we will denote in the following as RST. In the 1960s it was established that P1 and P2 are actually *unprovable propositions* within RST. As a result, we can either add P1 and P2 as axioms (thus obtaining a "standard" version of set theory), or we could add their negations, which results in the so-called non-Cantorian set theory. Although both of these possibilities produce logically consistent systems, the latter one is clearly more exotic and counterintuitive (much like non-Euclidean geometry).

To see why propositions P1 and P2 are unprovable in the framework of restricted set theory, we need to refer to some of Gödel's later work. Based on his classic results on the consistency and completeness of formal systems, Gödel introduced the class of *constructible sets* as a model for RST (in the following, we will denote this class by M). By definition, constructible sets are those that can be formed from simpler sets by a well defined sequence of steps. In this way, every statement of restricted set theory becomes "translatable" into a statement about M. Using such a model, Gödel was able to show that both P1 and P2 turn out to be *theorems* in M. As a result, he could legitimately claim that adding P1 and/or P2 as axioms in restricted set theory will *not* create consistency problems. If any such problems were to arise, their cause would necessarily be in restricted set theory. In this sense, Gödel's consistency proof was *relative* - he basically established that RST with P1 and P2 added as axioms is consistent if and only if RST itself is consistent.

The schematic diagram in Fig. 2.3 provides a further illustration of this point. Namely, if we were to assume that $\Phi(T_S) \subset T_M$ (where T_S and T_M denote the theorems of RST and M, respectively), this would obviously imply that RST is consistent. This property would remain unchanged if we were to add p_1 and p_2 to T_S, since both of these propositions map into T_M. Note, however, that $\Phi(T_S) \subset T_M$ is only an *assumption*, whose validity ultimately determines the truth of the overall statement. In that sense, the consistency proof is relative.

It can also be shown that adding the *negation* of P1 and P2 as axioms will result in relative consistency. In order to establish this fact, it is necessary (among

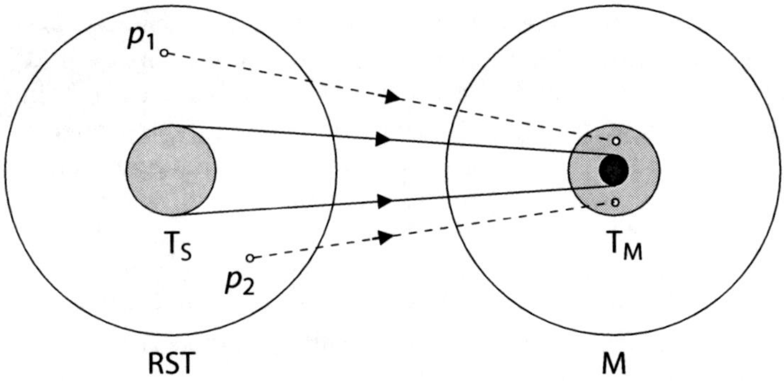

Fig. 2.3. An illustration of the relative consistency of standard set theory.

other things) to use a different model for restricted set theory. The details of the proof are quite complicated and will not be considered here.[16] For our purposes, it suffices to recognize that this result (combined with Gödel's conclusions) implies that P1 and P2 are *unprovable propositions* in the framework of RST. This provides yet another illustration of the incompleteness of formal mathematics, and of the interesting possibilities that can arise from this fact.

Non-Standard Analysis

We begin our brief description of this field with a sequence of straightforward propositions from standard mathematical analysis.

$$\begin{aligned} p_1 &: \quad \text{There exists a number } c \text{ such that } 0 < c < 1. \\ p_2 &: \quad \text{There exists a number } c \text{ such that } 0 < c < 1/2. \\ &\vdots \\ p_n &: \quad \text{There exists a number } c \text{ such that } 0 < c < 1/n. \\ &\vdots \end{aligned} \tag{2.53}$$

Clearly, any *finite* collection of these statements can be true simultaneously. However, the *whole set* cannot, since in standard analysis for any number c we can always find a sufficiently large n_0 such that $c \geq 1/n_0$. In other words, the proposition

$$p_\infty : \quad \text{For } \textit{any } n, \text{ there exists a number } c \text{ such that } 0 < c < 1/n.$$

is *false* in conventional mathematics.

The above result may perhaps be mildly counterintuitive, but there seems to be nothing particularly unusual about it. What is surprising, however, is the fact that there exists an "alternative" mathematical system in which p_∞ is actually *true*. To gain a better understanding of the properties of such a non-standard system, let us denote the set of all statements in standard mathematical analysis by M, and let L be a formal system into which we can "translate" these statements as appropriate strings of symbols. In this case, M obviously represents a *model* for L, in the sense proposed by Gödel. It is important to recognize in this context that not all statements of M can be properly formalized. In that respect, mathematics is surprisingly similar to human languages.

Every individual statement p_i in (2.53) can easily be "translated" into a string of symbols σ_i in L. The same holds true for any *finite subset* of statements p_i as well. It can be shown that under these conditions there exists another model for L, denoted M^*, in which the translations of σ_i are true both *individually* and as *a whole*. This result follows from the so-called Compactness Theorem, which is closely related to Gödel's work on completeness.

There is no doubt that M^* must contain some very strange objects in order for all σ_i to be true simultaneously. Among these objects are the elusive "infinitesimals", which are greater than zero and yet smaller than any ordinary positive number. Nothing like this exists in standard mathematics, as witnessed by the falsity of statement p_∞. An interesting question in this context is whether we can somehow relate M^* to M. It turns out that we can, since both are models for the *same* formal system L. Systems M and M^* are said to be *formally* equivalent, in the sense that there is a correspondence between those statements that are "translatable" into the formal system L. Of course, statements of M and M^* that are *not* translatable needn't be connected in any way. This situation is illustrated in Fig. 2.4, in which the shaded areas denote the formally equivalent parts of M and M^*.

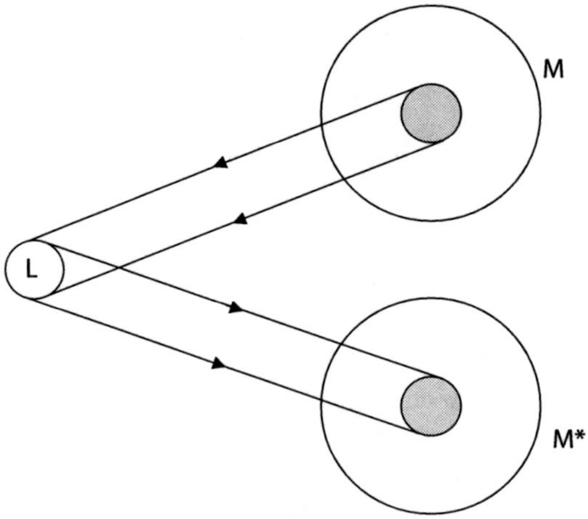

Fig. 2.4. The formally equivalent parts of M and M^*.

What does this equivalence really mean? Basically, it implies that what is true in M^* "translates" into what is true in M, via the common formal language L. As a result, it becomes possible to construct proofs in M^*, and then convert the result into statements about M. Most mathematicians require no more than that. They are usually not particularly interested in the "objective existence" of infinitesimals and related philosophical issues. Their primary concern is whether objects such as infinitesimals can be legitimately used in proofs, and the Compactness Theorem permits them to do so. This is very similar to the way in which complex numbers are used in electrical engineering. No one seriously ponders whether such numbers truly exist in the real world - all that matters is the fact that they greatly simplify the analysis of circuits with sinusoidal inputs.

Formal equivalence allows for some fascinating generalizations. For example, the true statement s_i: "10 is a positive number" in M can be mapped into L, and therefore has an equivalent statement s_i^* in M^*. This possibility (which is schematically illustrated in Fig. 2.5) leads to the conclusion that *all* standard numbers have a representation in M^*.

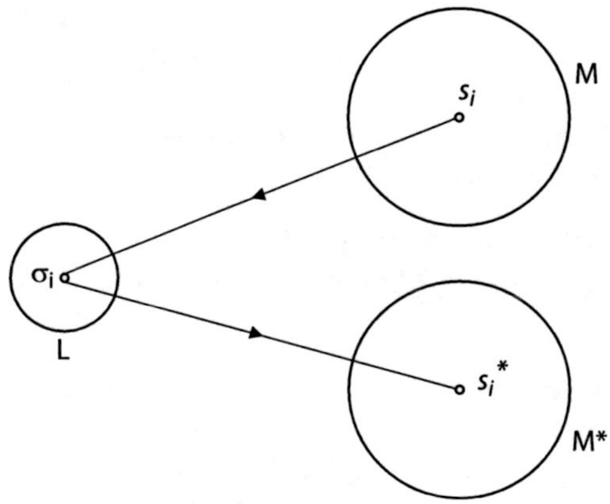

Fig. 2.5. Mapping of s_i into s_i^* via σ_i.

It is important to recognize that M^* also contains a variety of other objects such as infinitesimals and infinitely large numbers, which have no counterpart in M. Along these lines, we can think of any number $r \in M$ as having a "pseudo-real" equivalent $r^* \in M^*$, which is surrounded by a vast set of other pseudo-real objects. The entire infinitesimal "neighborhood" of r^* is sometimes referred to as a *monad*, following the philosophical terminology of Leibniz. For any number $\hat{r}^* = r^* + dr$ in such a monad, we say that r^* is its *standard part.*

Most mathematicians agree that these unusual generalizations of real numbers opened up a new and exciting line of research. However, there is no consensus as to where such an approach may ultimately lead. Whatever its future may be, it is important to remember that some key elements of this theory (such as the Compactness Theorem) directly relate to Gödel's results in the area of formal systems. One could therefore reasonably claim that Gödel paved the way for Abraham Robinson's pioneering work on infinitesimals, and provided the necessary conditions for the emergence of nonstandard analysis.[17]

2.5 Notes

1. This system was considered in: Douglas Hofstadter, *Gödel, Escher, Bach: An Eternal Golden Braid*, Basic Books, 1999.

2. Note that there could be many other theorems in the model system M which are *not* part of set W. We could think of these theorems as truths that cannot be "translated" into statements (*i.e.* well-formed strings) in S. In that respect, it is fair to say that the model M is "richer" than the original system.

3. A stronger version of Definition 2.4 is that $\Phi(T_S) = T_M$ for *all* possible interpretations (not just a particular model M). Such a definition of completeness is sometimes referred to as *adequacy*, or *completeness with respect to truth.*

4. More details on this system can be found in E. Nagel and J. R. Newman, *Gödel's Proof*, New York University Press, 2001.

5. The concept of a tautology, for example, does not arise within Sentential Calculus.

6. A rigorous proof can be found in: Kurt Gödel, *On Formally Undecidable Propositions of Principia Mathematica and Related Systems*, Dover Publications, 1992.

7. A considerable amount of literature has been devoted to Gödel's theorem and its proof. In this section, we will follow the approach proposed in Douglas Hofstadter's book: *Gödel, Escher Bach: An Eternal Golden Braid*, where this difficult topic is presented with remarkable clarity. The more ambitious reader can refer to Gödel's original paper for further details.

8. This table represents an abridged version of the one provided on page 268 of *Gödel, Escher Bach: An Eternal Golden Braid.*

9. The number theoretic interpretation of string (2.43) is $1 + 0 = 1$.

10. An excellent introductory exposition of the last two topics can be found in Davis and Hersh, *The Mathematical Experience*, Birkhauser, 1981.

11. Gregory Chaitin, *Meta-Math: The Quest for Omega*, Pantheon Books, 2005.

12. The fact that we are summing over self-delimiting programs is an important prerequisite for convergence. To see why this is so, suppose that the program represented by bits 1001 halts, and contributes 2^{-4} to the sum. In that case, we don't need to consider any other programs of the form $1001xxxxx$, because a self-delimiting program will *not* ask for more bits after halting. In this way we can eliminate a large number of redundant terms from the sum (2.51), and ensure that $0 \leq \Omega \leq 1$.

13. There are many other formulas for calculating π, some of which converge much faster than (2.52). A nice survey of these results can be found in: Borwein and Bailey, *Mathematics by Experiment*, A. K. Peters, 2004.

14. Gregory Chaitin, "Omega and why maths has no TOEs," *Plus Magazine*, **37**, December 2005.
15. Chaitin, *Meta-Math: The Quest for Omega.*
16. For further information, see: Paul Cohen, *Set Theory and the Continuum Hypothesis*, Addison-Wesley, 1966.
17. A classic text on this subject is: Abraham Robinson, *Nonstandard Analysis*, Princeton University Press, 1996.

Further Reading

Douglas R. Hofstadter, *Gödel, Escher, Bach: An Eternal Golden Braid*, Basic Books, 1999.

Ernest Nagel and James R. Newman, *Gödel's Proof*, New York University Press, 2001.

Raymond Smullyan, *Gödel's Incompleteness Theorems*, Oxford University Press, 1991.

Rebecca Goldstein, *Incompleteness: The Proof and Paradox of Kurt Gödel*, Norton W. W. & Co., 2005.

Torkel Franzén, *Gödel's Theorem: An Incomplete Guide to Its Use and Abuse*, A. K. Peters, 2005.

Gregory Chaitin, *Meta-Math: The Quest for Omega*, Pantheon Books, 2005.

Philip Davis and Reuben Hersh, *The Mathematical Experience*, Birkhauser, 1981.

Chapter 3

Quantum Mechanics

Much of what goes on in the world of quantum mechanics is difficult to describe in the language of classical physics. Concepts such as position, velocity or energy, for example, are still applicable, but must be used with great care in order to avoid erroneous interpretations. Although our mathematical models provide us with a great deal of information about the way microscopic particles behave, we must concede that many of the underlying ideas are highly counterintuitive, and are often best described using analogies and metaphors.

One of the most puzzling notions that we will encounter in this chapter is the so-called "state of superposition", which is used to describe unobserved particles. According to the Copenhagen interpretation of quantum mechanics, such particles exist as *collections of potentialities*, which are mutually exclusive in our everyday experience. Descriptions of this sort are clearly difficult to grasp, since they bring into question some of our most basic assumptions about our physical environment. We can easily envision, for example, an electron whose spin is oriented "up" or "down", but never both at once. And yet, that is precisely what quantum mechanics asks us to do. This paradoxical situation is resolved only after a measurement is made, at which point exactly one of the competing possibilities "materializes".

The state of superposition is formally described in terms of a complex-valued wave function $\psi(x,t)$, whose temporal evolution is governed by Schrödinger's equation. This equation is deterministic, and allows us to compute ψ at any point x and at any time t given a set of initial conditions. The problem, however, is that the wave function provides only partial information about the particle. Indeed, all that we can deduce from Schrödinger's equation are the *probabilities* of different outcomes, and we have no way of knowing what value will actually be recorded. This is not simply a reflection of our ignorance and inability to measure physical quantities at the microscopic level - it is an *irreducible uncertainty*, which is built into the very foundations of quantum mechanics. An obvious consequence of this property is that certain aspects of nature will necessarily remain unknowable to us.

The process by which a particle transitions from a state of superposition to a physically observable state poses a further challenge to our conventional understanding of "reality". This remains one of the most controversial issues in quantum mechanics, and is known as the "measurement problem." A key concept in this debate is the so-called "collapse" of the wave function, whose essence is still not fully understood (although it has been the subject of intense scrutiny for almost a century). Niels Bohr believed, for example, that a measurement device is a purely "classical" entity, while the particle itself conforms to the laws of quantum mechanics. From that perspective, the collapse of the wave function could be seen as a sort of "bridge", which somehow enables the particle to cross the gap that separates the microscopic domain from the classical one.

A very different response to this question is provided by the so-called "many worlds" interpretation of quantum mechanics, which was proposed by physicist Hugh Everett in the 1950s. On this view, there is no such thing as a "collapse", and *each* possible state supposedly materializes in a parallel universe. However strange it may sound, scientists take this theory seriously, since many of its predictions are consistent with existing experimental data.

Perhaps the most intriguing aspect of the measurement problem is the role of the observer. Unlike classical physics, which implicitly assumes that there exists an "external" world which can be studied objectively, quantum mechanics maintains that the physical attributes of a particle actually depend *on the way we choose to observe it.* We could decide, for example, to perform an experiment which would prove that an electron has wavelike characteristics (such as interference and spatial distribution). Alternatively, we could set up our experimental apparatus in a way that would clearly demonstrate the electron's corpuscular nature. Although these two results appear to be at odds with each other, modern physics claims that they are both legitimate, and represent equally valid manifestations of the electron's "physical reality." An electron, in other words, exhibits *particle–wave duality*, and the "face" that it will reveal to us depends on our method of investigation. The fact that the choice of measurement strategy is a *conscious* human decision further suggests that there may be no "real" physical world "out there" whose existence is completely independent of the observer.

Although we now have indisputable empirical evidence that particle-wave duality does indeed exist in nature, some aspects of this unusual property remain elusive. One of the most important unresolved questions in this context concerns the nature of the waves that are associated with particles. An explanation put forth by Max Born in the 1920s suggests that these waves represent time varying probabilities, which propagate through spacetime and interact much like ordinary light waves do. Whether these waves are "physically real" or are just a convenient mathematical tool for describing the particle's motion is still an open question, which cannot be settled empirically (since probabilities have no mass or energy).

A very different interpretation of the "wavelike" nature of matter was proposed by Richard Feynman in the 1960s. Feynman suggested that in moving between two points a quantum particle actually traverses *every* possible trajectory simultaneously. He then showed that the combined average over these paths produces the same probabilities as the ones calculated using the wave function. It is interesting to note that these two explanations (each of them counterintuitive in its own right) produce identical predictions, despite that fact they appear to contradict each other. As a result, scientists consider them to be equally acceptable.

Heisenberg's Uncertainty Principle places yet another restriction on what we can say about the microscopic world. Its most widely known version concerns the position and momentum of a particle (denoted in the following by x and p, respectively). If a particle happens to be in a state in which *neither* x *nor* p have a definite value, the probability distributions of these variables are characterized by their "spreads" Δp and Δx, respectively. What Heisenberg recognized was that Δp and Δx are *not* independent, and must satisfy the inequality $\Delta x \Delta p \geq \hbar/2$, where $\hbar$ is Planck's constant divided by 2π.

A common interpretation of this inequality is that we cannot simultaneously measure the position and velocity of a particle. While such a statement is technically correct, it fails to capture the essence of Heisenberg's result. What the Uncertainty Principle really tells us is that the more we know about the position of a particle, the less we can say about its momentum (and vice versa). Consider, for example, the extreme case when Δx is very small and Δp is very large. Under such circumstances, one could legitimately argue that the concept of momentum (which is a classical attribute) cannot be meaningfully associated with the particle, since all values of p are equally likely. This is the point where the language of Newtonian physics becomes inadequate (and can even be thoroughly misleading).

As a final point in this introductory section, we should also say a few words about the famous EPR paradox. Einstein was one of many scientists who rejected the view that quantum mechanics is inherently uncertain, and that a particle in a state of superposition is just a collection of potentialities. He suggested instead that the microscopic world conforms to a set of *deterministic* laws, whose precise nature has yet to be determined. In the 1930s, Einstein and his two graduate students, Podolski and Rosen, produced what they thought was a decisive argument against the Copenhagen interpretation of quantum theory. This argument, which became known as the EPR paradox, was the subject of heated debates for nearly fifty years. It was finally resolved in the early 1980s, when it was conclusively established that Einstein's conjecture was wrong.

The EPR paradox is best described in terms of an experiment in which an atom with spin zero disintegrates into two smaller particles (say A and B) that fly off in different directions. According to quantum mechanics, prior to a measurement the state of each particle ought to be a superposition of "up" and "down" spins. This theory also claims, however, that measuring a definite spin for particle A guarantees that we will subsequently measure the *opposite* spin for B, regardless of the distance between them (this is a direct consequence of the fact that the spin of the "parent" atom was zero).

Einstein argued that there is no physical mechanism that would allow information about measurements made on particle A to be "instantaneously transmitted" to particle B, given that no signal can propagate faster than light. Based on this apparent contradiction, he concluded that quantum mechanics is, in fact, an incomplete theory. He proposed an alternative explanation, according to which particles A and B were really subject to "hidden" deterministic laws, and are both in a state of *definite spin* which is decided at their "birth." Following this line of reasoning, he concluded that the behavior of such a system can always be described in terms "classical" probabilities.

Although Einstein and his collaborators were not aware of this at the time, there is actually a way to explicitly test the validity of their hypothesis. This test is based on a property of classical probabilities that was discovered by physicist John Bell in the early 1960s. This simple result proved to be crucial for verifying the validity of "hidden variable" theories, since it provided a condition that the particles must satisfy if their behavior is truly classical. The definitive experiment along these lines was conducted by Alain Aspect and his collaborators in 1981. Their measurements conclusively established that Bell's condition was *violated* in this case, which confirmed that quantum phenomena do *not* lend themselves to strictly deterministic descriptions.

Given such a result, the only consistent explanation of the EPR paradox is that no matter how far apart A and B are, their common "parentage" prevents us from treating them as separate entities. In technical terms, this means that the two particles shouldn't be described by separate wave functions, but rather by a *single* one (even if they end up in different galaxies). When particle A is observed, the *joint* wave function collapses, causing particle B to assume a definite spin value as well. According to this interpretation quantum mechanics must be viewed as a *non-local* theory, and all microscopic phenomena are somehow connected into a non-decomposable whole. Such a conclusion is truly counterintuitive, and has momentous consequences for the way we perceive physical reality. Among other things, it points to the fact that the "whole" can never be completely known by studying isolated parts, which places a significant restriction on what we can know about complex systems.

In the sections that follow, we will examine some these results more closely, and will introduce the mathematical formalism that is used in this context. Although our discussion will be limited to a relatively small number of topics, it should nevertheless provide the reader with a window into a fascinating and mysterious domain, which challenges the "obvious" truths of our everyday experience.

3.1 Quantum Operators

The state of a quantum particle is described by a complex-valued *wave function*, which is usually denoted by $\psi(x,t)$. The evolution of this function over time is governed by Schrödinger's equation

$$i\hbar\frac{\partial\psi}{\partial t} = \hat{H}\psi \tag{3.1}$$

where $\hat{H}$ represents a linear operator, and $\hbar$ is Planck's constant divided by 2π. For a single particle of mass m in an external field $V(x,y,z)$, $\hat{H}$ has the form

$$\hat{H} = -\frac{\hbar^2}{2m}\nabla^2 + V(x, y, z) \tag{3.2}$$

where

$$\nabla^2 = \frac{\partial^2}{\partial x^2} + \frac{\partial^2}{\partial y^2} + \frac{\partial^2}{\partial z^2} \tag{3.3}$$

There is an interesting connection between operator $\hat{H}$ and the total energy of a classical particle. In the classical case, the sum of the kinetic and potential energy can be expressed as

$$E = \frac{1}{2}m(v_x^2 + v_y^2 + v_z^2) + V(x, y, z) \tag{3.4}$$

or equivalently (in terms of momentum) as

$$E = \frac{1}{2m}(p_x^2 + p_y^2 + p_z^2) + V(x, y, z) \tag{3.5}$$

If the momentum components in (3.5) are now substituted by *operators*

$$\hat{p}_x = i\hbar\frac{\partial}{\partial x}; \qquad \hat{p}_y = i\hbar\frac{\partial}{\partial y}; \qquad \hat{p}_z = i\hbar\frac{\partial}{\partial z} \tag{3.6}$$

we obtain

$$\frac{1}{2m}(\hat{p}_x^2 + \hat{p}_y^2 + \hat{p}_z^2) = -\frac{\hbar^2}{2m}\nabla^2 \tag{3.7}$$

In other words, replacing p_x, p_y and p_z in (3.5) with their operator counterparts directly produces $\hat{H}$.

What does the wave function $\psi(x,t)$ actually tell us? That depends to some extent on what it is that we want to measure. Suppose, for example, that we are interested in an observable quantity q. With any such quantity, we can associate an operator $\hat{Q}$, much like we did with the momentum operator in (3.6). If we consider only stationary states, the eigenfunctions[1] of $\hat{Q}$ will depend on x only, satisfying

$$\hat{Q}\phi_i(x) = q_i\phi_i(x) \tag{3.8}$$

In equation (3.8), real numbers q_i $(i = 1, 2, \ldots)$ represent the *eigenvalues* that correspond to functions $\phi_i(x)$ $(i = 1, 2, \ldots)$. The quantum mechanical interpretation of this equation is that the particle is in a *definite* state of q only when $\psi(x) \in \{\phi_1(x), \phi_2(x), \ldots\}$. In other words, if $\psi(x) = \phi_i(x)$, a measurement will certainly produce $q = q_i$.

What can we say about quantum states that do *not* correspond to eigenfunctions? Before answering this question, we should first note that operator $\hat{Q}$ is always chosen so that it satisfies the following three properties:

Property 1. All eigenvalues of $\hat{Q}$ must be *real numbers.* This is important because it allows us to identify measurable physical quantities with eigenvalues.

Property 2. The eigenfunctions of $\hat{Q}$ form a *basis* in the function space. This means that *any* function $f(x)$ can be expressed in terms of $\{\phi_1(x), \phi_2(x), \ldots\}$ as

$$f(x) = \sum_i \beta_i \phi_i(x) \tag{3.9}$$

where β_i $(i = 1, 2, \ldots)$ are constant coefficients.

Property 3. The *scalar product* [2] of two eigenfunctions, which is defined as

$$\langle \phi_i(x), \phi_j(x) \rangle = \int \phi_i^*(\rho)\phi_j(\rho)d\rho \tag{3.10}$$

satisfies

$$\langle \phi_i(x), \phi_j(x) \rangle = \begin{cases} 1, & i = j \\ 0, & i \neq j \end{cases} \tag{3.11}$$

When this property holds, the basis $\{\phi_1(x), \phi_2(x), \ldots\}$ is said to be *orthonormal.*

The existence of an orthonormal basis for $\hat{Q}$ allows us to express the wave function $\psi(x)$ as

$$\psi(x) = \sum_i \alpha_i(q_i)\phi_i(x) \tag{3.12}$$

In such a case, the particle is *not* in any definite state of q, but is rather a *superposition* of all observable states. What this means is that a measurement could produce *any* of the possible values for q, but we cannot tell in advance which one. What we do know, however, is the probability of measuring a particular value q_i. This probability is directly related to the coefficients $\alpha_i(q_i)$, and can be expressed as

$$P(q_i) = |\alpha_i(q_i)|^2 \tag{3.13}$$

To see how $P(q_i)$ can be computed, we should recall that the scalar product defined in (3.10) satisfies [3]

$$\langle \sum_i a_i f_i(x), g(x) \rangle = \sum_i a_i^* \langle f_i(x), g(x) \rangle \tag{3.14}$$

where $f_i(x)$ and $g(x)$ are arbitrary complex-valued functions. We can then use (3.11) and (3.12) to obtain

$$\langle \psi(x), \phi_i(x) \rangle = \sum_j \alpha_j^*(q_j) \langle \phi_j(x), \phi_i(x) \rangle = \alpha_i^*(q_i) \tag{3.15}$$

which allows us to express the probability as

$$P(q_i) = |\alpha_i(q_i)|^2 = |\langle \psi(x), \phi_i(x) \rangle|^2. \tag{3.16}$$

Remark 3.1. In order to treat the terms $|\alpha_i(q_i)|^2$ as probabilities, we must ensure that

$$\sum_i |\alpha_i(q_i)|^2 = 1 \tag{3.17}$$

We can do this by using *normalized* wave functions, which must satisfy the constraint $\langle \psi(x), \psi(x) \rangle = 1$. To see why this is so, it suffices to observe that

$$\begin{aligned} \langle \psi(x), \psi(x) \rangle &= \langle \sum_i \alpha_i \phi_i(x), \sum_j \alpha_j \phi_j(x) \rangle = \\ &= \sum_{i,j} \alpha_i^* \alpha_j \langle \phi_i(x), \phi_j(x) \rangle = \sum_i |\alpha_i(q_i)|^2 \end{aligned} \tag{3.18}$$

What happens when we perform an actual measurement? In that case, the wave function *collapses* into exactly one of the eigenfunctions $\phi_i(x)$, and automatically moves into a state that corresponds to eigenvalue q_i. It is important to recognize that all subsequent measurements will produce the *same* result. Indeed, since the system is in state $\psi(x) = \phi_i(x)$ after the initial measurement, the probability of detecting some $q_j \neq q_i$ is

$$P(q_j) = |\langle \phi_i(x), \phi_j(x) \rangle|^2 = 0 \tag{3.19}$$

Without this property (which follows from the orthogonality condition (3.11)), our predictions about quantum phenomena would be completely unreliable.

The Continuous Spectrum

In the preceding discussion, we assumed that operator $\hat{Q}$ has a *discrete* set of eigenfunctions and eigenvalues. But what happens if the possible outcomes represent a *continuum*? In such cases, the expansion (3.12) turns into an *integral*

$$\psi(x) = \int \alpha(q) \phi(x, q) dq, \tag{3.20}$$

much like Fourier series turn into Fourier transforms when the function is aperiodic. The quantity $|\alpha(q)|^2$ can be interpreted as the *probability density*, where

$$dP = |\alpha(q)|^2 dq \tag{3.21}$$

represents the infinitesimal probability that the outcome will fall in the interval $[q, q + dq]$. It is not difficult to show that

$$\alpha^*(q) = \langle \psi(x), \phi(x, q) \rangle = \int \psi^*(x) \phi(x, q) dx \tag{3.22}$$

which is a continuous version of expression (3.15).

Example 3.1. In the one-dimensional case, the quantum *momentum operator* takes the form

$$\hat{p} = i\hbar \frac{\partial}{\partial x} \tag{3.23}$$

States of definite momentum correspond to the eigenfunctions $\phi(x, p)$ of this operator, which must satisfy

$$i\hbar \frac{\partial \phi(x, p)}{\partial x} = p\, \phi(x, p) \tag{3.24}$$

It is easily verified that functions of the form

$$\phi(x,p) = e^{-ipx/\hbar} \tag{3.25}$$

are solutions of (3.24) for *any* choice of p. This indicates that the momentum operator has a *continuous* set of eigenvalues.[4]

Following (3.20), we can express the quantum state $\psi(x)$ in terms of momentum components as

$$\psi(x) = \int \alpha(p)\phi(x,p)dp = \int \alpha(p)e^{-ipx/\hbar}dp \tag{3.26}$$

As noted earlier, in such a representation $|\alpha(p)|^2 dp$ corresponds to the probability that the measurement will produce a momentum value in the interval $[p, p+dp]$.

Example 3.2. The one-dimensional *position operator* is simply a multiplication by x, whose eigenfunctions must satisfy

$$\hat{X}\varphi(x,\rho) = x\varphi(x,\rho) = \rho\varphi(x,\rho) \tag{3.27}$$

In (3.27), $\varphi(x,\rho)$ represents a state of definite position, in which $x = \rho$. It is not difficult to verify that $\varphi(x,\rho) = \delta(x-\rho)$, since Dirac's delta function has the property

$$(x-\rho)\delta(x-\rho) = 0 \tag{3.28}$$

With that in mind, we can express $\psi(x)$ in terms of states with definite position as

$$\psi(x) = \int \alpha(\rho)\delta(x-\rho)d\rho \tag{3.29}$$

where $|\alpha(\rho)|^2$ represents the probability density for measuring $x = \rho$.

This result has an interesting interpretation, if we recall that Dirac's delta function satisfies

$$\int \alpha(\rho)\delta(x-\rho)d\rho = \alpha(x) \tag{3.30}$$

by definition. Comparing (3.29) and (3.30), it follows that $\psi(x) = \alpha(x)$, which allows us to view $|\psi(x)|^2$ as the probability density that a measurement will find the particle in position x.

In cases when a continuum of outcomes is possible, the probability of observing a particular value $q \in [q_0, q_0 + dq]$ is an infinitely small number. As a result, it often makes more sense to ask whether quantity q falls in a certain interval $W = [q_a, q_b]$. In order to make such an evaluation, we can rewrite (3.20) as

$$\psi(x) = \int\limits_{q\in W} \alpha(q)\phi(x,q)dq + \int\limits_{q\notin W} \alpha(q)\phi(x,q)dq = \psi_A(x) + \psi_B(x) \tag{3.31}$$

It is important to note that although functions $\psi_A(x)$ and $\psi_B(x)$ are guaranteed to be orthogonal (that is, $\langle\psi_A(x), \psi_B(x)\rangle = 0$), they are *not normalized.* Indeed, since

$$\begin{aligned}\langle\psi(x), \psi(x)\rangle &= \langle\psi_A(x) + \psi_B(x), \psi_A(x) + \psi_B(x)\rangle = \\ &= \langle\psi_A(x), \psi_A(x)\rangle + \langle\psi_B(x), \psi_B(x)\rangle = 1\end{aligned} \tag{3.32}$$

we have that $\langle \psi_A(x), \psi_A(x) \rangle = a^2$ and $\langle \psi_B(x), \psi_B(x) \rangle = b^2$, with $a^2 + b^2 = 1$. We can eliminate this problem by normalizing $\psi_A(x)$ and $\psi_B(x)$ as

$$\bar{\psi}_A(x) = \frac{1}{a}\psi_A(x); \qquad \bar{\psi}_B(x) = \frac{1}{b}\psi_B(x) \tag{3.33}$$

and rewriting $\psi(x)$ as

$$\psi(x) = a\bar{\psi}_A(x) + b\bar{\psi}_B(x) \tag{3.34}$$

In this representation, a^2 is the probability that $\psi(x)$ will collapse into state $\bar{\psi}_A(x)$, which corresponds to $q_a \leq q \leq q_b$. Note that $\bar{\psi}_A(x)$ is *not* a state with a definite value of q, but rather corresponds to a *fixed range* of possible values.

3.2 Heisenberg's Uncertainty Principle

Heisenberg's Uncertainty Principle is one of the best known results of quantum mechanics. In simple terms, this principle maintains that we cannot simultaneously detect the position and momentum of a quantum particle without an error. The problem that we are dealing with here is not unlike measuring the distance between two cities. If cities were points, this distance could be determined with arbitrary precision. However, since they actually have finite dimensions, it is impossible to reduce the measurement error below a certain threshold.

In order to provide a more formal interpretation of this principle, we must first establish several preliminary results. We begin with the following definition.

Definition 3.1. We will say that operators $\hat{A}$ and $\hat{B}$ *commute* if they satisfy

$$\hat{A}\hat{B}\psi = \hat{B}\hat{A}\psi \tag{3.35}$$

for any function $\psi(x)$. This property is formally expressed as

$$[\hat{A}, \hat{B}] \equiv \hat{A}\hat{B} - \hat{B}\hat{A} = 0 \tag{3.36}$$

Proposition 3.1. Any two operators that commute have a common set of eigenfunctions.

Proof. Suppose that $\hat{A}$ and $\hat{B}$ commute, and let ψ_n be an eigenfunction of $\hat{B}$. Then, there exists a number b_n such that

$$\hat{B}\psi_n = b_n\psi_n \tag{3.37}$$

It is not difficult to show that $\varphi_n \equiv \hat{A}\psi_n$ is also an eigenfunction of $\hat{B}$. Indeed, since $\hat{A}\hat{B} = \hat{B}\hat{A}$, we have

$$\hat{B}\varphi_n = \hat{B}\hat{A}\psi_n = \hat{A}\hat{B}\psi_n = b_n\hat{A}\psi_n = b_n\varphi_n \tag{3.38}$$

Observing that φ_n and ψ_n are eigenfunctions of $\hat{B}$ that correspond to the *same* eigenvalue b_n, it follows that they can differ at most by a constant factor α_n (*i.e.* $\varphi_n = \alpha_n\psi_n$). Consequently,

$$\hat{A}\psi_n \equiv \varphi_n = \alpha_n\psi_n \tag{3.39}$$

which means that ψ_n is an eigenfunction of $\hat{A}$ as well.

Remark 3.2. Since eigenfunctions are associated with observable states, Proposition 3.1 tells us that we can *simultaneously* measure quantities A and B whenever the corresponding operators commute. It is important to recognize, however, that commutation is *not* a universal property in quantum mechanics. Operators such as $\hat{x}$ and $\hat{p}$, for example, satisfy

$$[\hat{x}, \hat{p}] = i\hbar \tag{3.40}$$

which means that the states of definite momentum and position are *mutually exclusive*.

Proposition 3.2. Let us assume that $\hat{A}$ is an operator that corresponds to some observable quantity A. The *expected value* of A (denoted $\langle A\rangle$), and its *dispersion* ΔA can then be expressed as

$$\langle A\rangle = \langle \psi, \hat{A}\psi\rangle \tag{3.41}$$

and

$$(\Delta A)^2 \equiv \langle A^2\rangle - \langle A\rangle^2 = \langle \psi, (\Delta\hat{A})^2\psi\rangle \tag{3.42}$$

where operator $\Delta\hat{A}$ is defined as

$$\Delta\hat{A} \equiv \hat{A} - \langle A\rangle \tag{3.43}$$

Proof. We will demonstrate this property for the case when $\hat{A}$ has discrete eigenvalues $\{a_1, a_2, \ldots\}$, with corresponding eigenfunctions $\{\phi_1, \phi_2, \ldots\}$ (the proof for the continuous case is similar). We begin by observing that expansion (3.12) and property (3.8) allow us to write

$$\begin{aligned} \langle \psi(x), \hat{A}\psi(x)\rangle &= \langle \sum_i \alpha_i\phi_i(x), \sum_j \alpha_j a_j \phi_j(x)\rangle = \\ &= \sum_{i,j} \alpha_i^* \alpha_j a_j \langle \phi_i(x), \phi_j(x)\rangle = \sum_i a_i|\alpha_i(a_i)|^2 \end{aligned} \tag{3.44}$$

Since $|\alpha_i(a_i)|^2$ is the probability of observing a_i, (3.44) can be interpreted as the *expected value* of A.

Regarding the dispersion, it suffices to recall the definition of operator $\Delta\hat{A}$, which implies that

$$(\Delta\hat{A})^2 = \hat{A}^2 - 2\hat{A}\langle A\rangle + \langle A\rangle^2 \tag{3.45}$$

From this expression, we directly obtain

$$\langle \psi, (\Delta\hat{A})^2\psi\rangle = \langle \psi, \hat{A}^2\psi\rangle - 2\langle A\rangle\langle \psi, \hat{A}\psi\rangle + \langle A\rangle^2 = \langle A^2\rangle - \langle A\rangle^2 \tag{3.46}$$

Remark 3.3. Note that the dispersion ΔA is in fact the quantum equivalent of the standard deviation. This allows us to interpret ΔA as the predicted statistical spread of results when repeated measurements of A are made.

Heisenberg's Principle is usually formulated in terms of the dispersions of the position and momentum, as

$$\Delta x \Delta p \geq \frac{\hbar}{2} \tag{3.47}$$

In order to derive this relationship, let us assume that operators $\Delta\hat{A} = \hat{A} - \langle A\rangle$ and $\Delta\hat{B} = \hat{B} - \langle B\rangle$ correspond to the dispersions of some arbitrary quantities A and B. It is not difficult to see that inequality

$$J(\eta) = \int |(\eta\Delta\hat{A} - i\Delta\hat{B})\psi(x)|^2 dx \geq 0 \tag{3.48}$$

holds for *any* real number η. Using the fact that $[\Delta\hat{A}, \Delta\hat{B}] = [\hat{A}, \hat{B}]$, this integral can be expressed as

$$J(\eta) = \eta^2 \int \psi^*(\Delta\hat{A})^2\psi dx - i\eta \int \psi^*[\hat{A}, \hat{B}]\psi dx + \int \psi^*(\Delta\hat{B})^2\psi dx \tag{3.49}$$

In the special case when $\hat{A} = \hat{x}$ and $\hat{B} = \hat{p}$, the term $[\hat{A}, \hat{B}]$ in the second integral becomes $[\hat{A}, \hat{B}] = [\hat{x}, \hat{p}] = i\hbar$. Recalling (3.42), $J(\eta)$ now takes the form

$$J(\eta) = \eta^2(\Delta x)^2 + \eta\hbar + (\Delta p)^2 \geq 0 \tag{3.50}$$

Since this is a quadratic inequality in η, it is easily verified that (3.50) is satisfied if and only if (3.47) holds.

Example 3.3. As an illustration of Heisenberg's Principle, let us consider the case when $\psi(x)$ has the form of a *wave packet*

$$\psi(x) = \frac{1}{(2\pi)^{1/4}\sqrt{\Delta x}} \exp\left[\frac{i\langle p\rangle x}{\hbar} - \frac{x^2}{4(\Delta x)^2}\right] \tag{3.51}$$

where $\langle p\rangle$ denotes the expected value of the momentum ($\langle x\rangle$ is assumed to be zero for simplicity). It is easily recognized that in this case

$$|\psi(x)|^2 = \frac{1}{\sqrt{2\pi}\Delta x} \exp\left[-\frac{x^2}{2(\Delta x)^2}\right] \tag{3.52}$$

represents a Gaussian distribution, whose standard deviation Δx corresponds to the dispersion of the position. The effect of Δx on the wave function is illustrated in Figs. 3.1 and 3.2, which were obtained for $\Delta x = 1$ and $\Delta x = 0.01$, respectively (with $\langle p\rangle = 7\hbar$ in both cases). Fig. 3.2 indicates that when $\Delta x \to 0$ the probability density $|\psi(x)|^2$ approaches $\delta(x)$, which corresponds to a state of *definite position.*

Using the fact that

$$\alpha^*(p) = \int \psi^*(x)\phi(x, p)dx \tag{3.53}$$

and recalling (3.25), we can express the probability density of the momentum as

$$|\alpha(p)|^2 = A \exp\left[-\frac{2(\Delta x)^2(p - \langle p\rangle)^2}{\hbar^2}\right] \tag{3.54}$$

where A is a constant. This is a Gaussian distribution as well, with a standard deviation

$$\Delta p = \frac{\hbar}{2(\Delta x)} \tag{3.55}$$

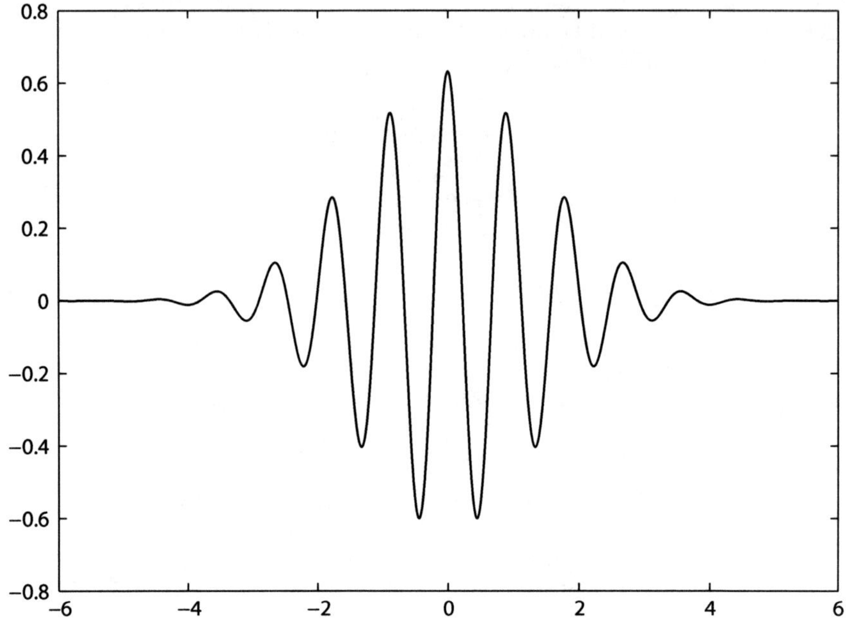

Fig. 3.1. The real part of $\psi(x)$ for $\Delta x = 1$.

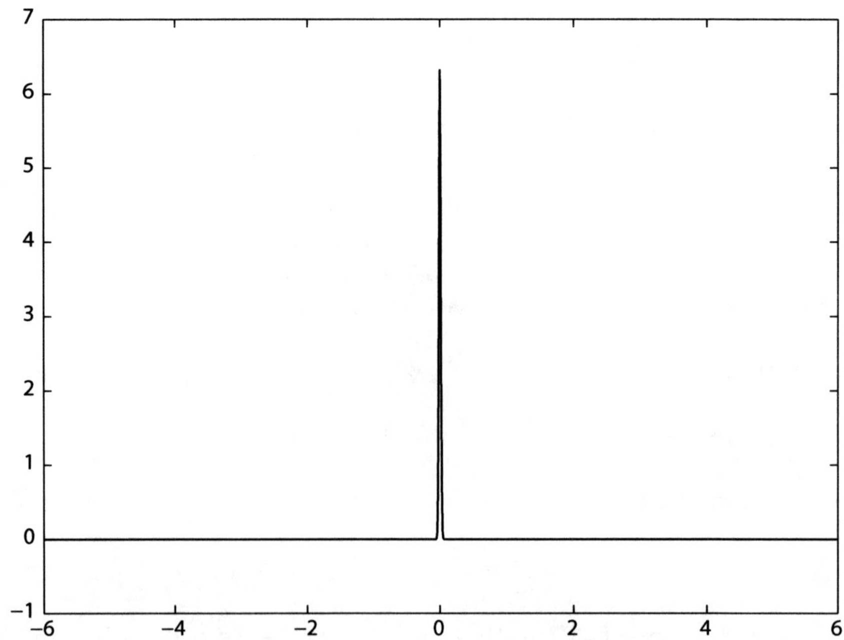

Fig. 3.2. The real part of $\psi(x)$ for $\Delta x = 0.01$.

Since Δp represents the dispersion of the momentum, it is now easily verified that

$$\Delta x \Delta p = \frac{\hbar}{2} \tag{3.56}$$

which is actually the limiting case of Heisenberg's result.

Example 3.4. What would happen if we were to know the *exact* location of the particle (let us denote this location by x_0)? For this to be possible, the particle must be in a state of *definite position*, which implies

$$\psi(x) = \delta(x - x_0) \tag{3.57}$$

Substituting this into (3.53), we obtain

$$\begin{aligned} \alpha^*(p) &= \int \delta(x - x_0)\phi(x, p)dx = \\ &= \int \delta(x - x_0)e^{-ipx/\hbar}dx = e^{-ipx_0/\hbar} \end{aligned} \tag{3.58}$$

Since

$$|\alpha(p)|^2 = |e^{-ipx_0/\hbar}| = 1 \tag{3.59}$$

it follows that the probability density is *exactly the same* for all p. In other words, if the *precise* position of the particle is known, its momentum turns out to be *completely uncertain.*

Heisenberg's principle has been viewed by many as a theoretical proof that certain properties of quantum particles are inherently unknowable to us. We should bear in mind, however, that this is not just a matter of our inability to simultaneously observe the exact position and velocity of a particle. What Heisenberg's principle really tells us is that states of definite position and momentum *cannot* coexist. Thus, if we determine the position precisely, there is *no definite momentum to be measured.*

It is also important to recognize that the uncertainty principle is not limited to momentum and position. One of its most important variants can be formulated as

$$\Delta E \Delta t \geq \frac{\hbar}{2} \tag{3.60}$$

where ΔE represents the most likely change in energy, and Δt is the time over which this change takes place. Inequality (3.60) suggests that over very short periods of time energy can fluctuate dramatically, even to the point of producing a short lived particle-antiparticle pair in empty space (according to special relativity, pure energy can be converted into matter and vice versa). One of the consequences of this fact is that "vacuum" is actually teeming with microscopic activity, and should be thought of as empty only *on the average.* The existence of this microscopic "quantum frenzy" has far reaching implications, and is one of the main obstacles to the development of a consistent theory of quantum gravity.

3.3 The EPR Paradox and Quantum Entanglement

In the 1930s, Einstein and his two graduate students, Podolski and Rosen, produced what they thought was a decisive argument against the so-called Copenhagen interpretation of quantum mechanics, which maintains that the microscopic world is inherently uncertain. This argument, which became known as the EPR paradox, was the subject of heated debates for nearly five decades decades. When the question was finally resolved in the early 1980s, it became clear that Einstein's premise was wrong, and that the behavior of quantum particles cannot be explained by deterministic laws and "hidden variables".

In order to understand the nature of this paradox, we first need to briefly describe what is meant by the "spin" of a particle. The notion of spin arises when relativistic effects are incorporated into quantum mechanics. In order to gain an intuitive grasp of this concept, it is helpful to imagine a ball with a unit radius that spins around its own axis. All points that belong to such a ball obviously experience circular motion (except, of course, those that are *on* the axis itself). Given the angular velocity of this object, classical physics allows us to uniquely determine the corresponding angular momentum.

The problem with extending this concept to quantum mechanics stems from the fact that particles cannot be viewed as microscopic "balls." Instead, they are formally represented as *point-like* entities which have no physical dimensions. Such objects clearly cannot rotate around their own axis in any conventional sense, and it is hard to see how one could associate an angular momentum with them. And yet, that is precisely what quantum mechanics does when it talks about the "spin" of a particle.

In dealing with quantum spin, it is important to keep in mind that this property has no counterpart in the world of our everyday experience. We can perhaps interpret it as something that is mathematically similar to conventional rotation, but is ultimately different in kind. The nature of this difference is clearly illustrated by the fact that spin is an *intrinsic* characteristic of the particle, just like its mass or charge. This is quite unlike classical physics, where the angular momentum explicitly depends on the velocity of the object.

Spin can be characterized by two pieces of information:

1. *The amount of spin*, which is fixed for a given particle. Particles whose spin is an odd multiple of $\hbar/2$ are referred to as *fermions*, while those whose spin is an integer multiple of $\hbar$ (including zero) are know as *bosons*.

2. *The direction of the spin*, which can be clockwise or counter clockwise. These two possibilities correspond to 'spin up' and 'spin down' states, respectively.

To get a sense for how quantum spin is treated mathematically, let us consider an isolated electron, which has the smallest possible amount of spin for a fermion (that is, $\hbar/2$). What we don't know about the electron's spin is its *direction*, and this is something that we can measure. Since there are only two possible outcomes, the states of *definite spin* can be described by wave functions ψ_1 and ψ_2. If the electron is *not* in one of these two states, its wave function can be expressed as their superposition

$$\psi = \alpha_1\psi_1 + \alpha_2\psi_2 \tag{3.61}$$

where $|\alpha_1|^2$ and $|\alpha_2|^2$ represent the probabilities of measuring 'spin up' and 'spin down', respectively.

This brief (and admittedly superficial) discussion of quantum spin is all that we really need in order to understand the EPR paradox. The essence of this paradox is perhaps best described in terms of an experiment in which an atom with spin zero disintegrates into two smaller particles (say A and B) that fly off in different directions. According to quantum mechanics, prior to a measurement the state of each particle ought to be a superposition of "up" and "down" spins. Note, however, that if we happen to measure spin "up" for particle A, it is guaranteed that we will subsequently measure spin "down" for B, regardless of the distance between them. This is due to the conservation of angular momentum, which holds both in classical and quantum physics (keep in mind that in this case the spin of the "parent" atom was zero).

The EPR paradox relates to the fact that measuring the spin of particle A somehow places particle B in a state of definite spin as well, although B itself has not yet been "disturbed" (B could actually be very far from A when the measurement occurs). There is no physical mechanism that would allow information about measurements made on particle A to be "instantaneously transmitted" to particle B, given that no signal can propagate faster than light. Based on this apparent contradiction, Einstein argued that quantum mechanics is, in fact, an incomplete theory. He suggested that particles A and B were really subject to "hidden" deterministic laws, and are both in a state of *definite spin* which is decided at their "birth." He further hypothesized that we may not be able to predict exactly what this state will be, but we can always assign a "classical" probability to each outcome.

Although Einstein and his collaborators were not aware of this at the time, there is actually a way to explicitly test the validity of their conjecture. This test is based on certain properties of classical probabilities which were discovered by physicist John Bell in 1964. To get a sense for what Bell's result means, suppose that particles A and B have *definite* spins, and consider all possible combinations in the three spatial directions. In Table 3.1, these states are summarized as triplets (s_x, s_y, s_z), which are necessarily complementary for the two particles along each axis, due to the conservation of momentum.

As an illustration, let us consider the probability that particle A has spin 'up' in the x-direction (*i.e.* $x\uparrow$) and that B has spin 'up' in the y-direction (*i.e.* $y\uparrow$). Out of the eight possible scenarios in the table, only combinations 3 and 4 correspond to this particular outcome. Consequently, the overall probability $P(x\uparrow, y\uparrow)$ can be expressed as

$$P(x\uparrow, y\uparrow) = \frac{n_3 + n_4}{N} \tag{3.62}$$

	Particle A	Particle B
1	$(x\uparrow, y\uparrow, z\uparrow)$	$(x\downarrow, y\downarrow, z\downarrow)$
2	$(x\uparrow, y\uparrow, z\downarrow)$	$(x\downarrow, y\downarrow, z\uparrow)$
3	$(x\uparrow, y\downarrow, z\uparrow)$	$(x\downarrow, y\uparrow, z\downarrow)$
4	$(x\uparrow, y\downarrow, z\downarrow)$	$(x\downarrow, y\uparrow, z\uparrow)$
5	$(x\downarrow, y\uparrow, z\uparrow)$	$(x\uparrow, y\downarrow, z\downarrow)$
6	$(x\downarrow, y\uparrow, z\downarrow)$	$(x\uparrow, y\downarrow, z\uparrow)$
7	$(x\downarrow, y\downarrow, z\uparrow)$	$(x\uparrow, y\uparrow, z\downarrow)$
8	$(x\downarrow, y\downarrow, z\downarrow)$	$(x\uparrow, y\uparrow, z\uparrow)$

Table 3.1. Possible spin combinations.

where n_3 and n_4 represent the number of generated particles with combinations 3 and 4, respectively, while N denotes the total number of particles. Following the same procedure, probabilities $P(x\uparrow, z\uparrow)$ and $P(z\uparrow, y\uparrow)$ can be computed as

$$P(x\uparrow, z\uparrow) = \frac{n_2 + n_4}{N} \tag{3.63}$$

and

$$P(z\uparrow, y\uparrow) = \frac{n_3 + n_7}{N} \tag{3.64}$$

Observing that

$$\frac{n_3 + n_4}{N} < \frac{n_2 + n_4}{N} + \frac{n_3 + n_7}{N} \tag{3.65}$$

we now have that

$$P(x\uparrow, y\uparrow) < P(x\uparrow, z\uparrow) + P(z\uparrow, y\uparrow) \tag{3.66}$$

which is one of several possible formulations of *Bell's inequality.*

This simple result proved to be crucial for verifying the validity of "hidden variable" theories, since it provided a condition that the particles must satisfy if their behavior is truly classical. The definitive experiment along these lines was conducted by Alain Aspect and his collaborators in 1981. Their measurements conclusively established that Bell's inequality was *violated* in this case, which confirmed that quantum phenomena do *not* lend themselves to strictly deterministic descriptions.

Given such a result, the only consistent explanation of the EPR paradox is that no matter how far apart A and B are, their common "parentage" prevents us from treating them as separate entities. In technical terms, this means that the two particles shouldn't be described by separate wave functions, but rather by a *single* one (even if they end up in different galaxies). When particle A is observed, the *joint* wave function collapses, causing particle B to assume a definite spin value as well. According to this interpretation quantum mechanics must be viewed as a *non-local* theory, and all microscopic phenomena are somehow connected into a non-decomposable whole.

Quantum Entanglement

The process that produces a "joint" wave function for a collection of interacting particles is known as *quantum entanglement.* In order to understand what this means, let us first consider a classical system S, which consists of two subsystems S_1 and S_2. If the possible states of S_1 and S_2 are $\{x_1,\ x_2,\ x_3,\ldots\}$ and $\{y_1,\ y_2,\ y_3,\ldots\}$, respectively, the overall state of the system can always be uniquely associated with a pair of "microstates" (x_i, y_j). In this way, the behavior of the "whole" can always be reduced to the behavior of its constituent parts.

In quantum mechanics, the situation is considerably more complicated. To see why this is so, suppose that subsystem S_1 is initially in state ψ_i, which corresponds to a definite value of some physical quantity A. We will make a similar assumption for system S_2 as well, the only difference being that its state ϕ_j corresponds to a definite value of physical quantity B. When these two subsystems *interact*, the state of the resulting composite system S can be described as

$$\chi = \sum_{i,\ j} \alpha_{ij}(\psi_i \otimes \phi_j) \tag{3.67}$$

where $\otimes$ denotes a mathematical operation that combines functions ψ_i and ϕ_j in a special way.

An important property of this operation it that the state χ of the overall system *isn't* always reducible to the states of S_1 and S_2. As a result, the two subsystems must be viewed as elements of a greater "whole" whose dynamics *include* and *transcend* the dynamics of its parts. The following example illustrates how such "entangled" states can arise in a simple hypothetical model.

Example 3.5. Let us consider two subsystems S_1 and S_2 whose individual states are described in terms of vectors

$$x = \begin{bmatrix} x_1 \\ x_2 \\ x_3 \\ x_4 \end{bmatrix} \tag{3.68}$$

and

$$y = \begin{bmatrix} y_1 \\ y_2 \\ y_3 \end{bmatrix}, \tag{3.69}$$

respectively. It is easily verified that any state of S_1 can be expressed as

$$x = x_1\psi_1 + x_2\psi_2 + x_3\psi_3 + x_4\psi_4 \tag{3.70}$$

where vectors

$$\psi_1 = \begin{bmatrix} 1 \\ 0 \\ 0 \\ 0 \end{bmatrix}; \quad \psi_2 = \begin{bmatrix} 0 \\ 1 \\ 0 \\ 0 \end{bmatrix}; \quad \psi_3 = \begin{bmatrix} 0 \\ 0 \\ 1 \\ 0 \end{bmatrix}; \quad \psi_4 = \begin{bmatrix} 0 \\ 0 \\ 0 \\ 1 \end{bmatrix} \tag{3.71}$$

represent an orthonormal basis in R^4. The same holds true for subsystem S_2, whose states can be expanded as

$$y = y_1\phi_1 + y_2\phi_2 + y_3\phi_3 \tag{3.72}$$

using

$$\phi_1 = \begin{bmatrix} 1 \\ 0 \\ 0 \end{bmatrix}; \quad \phi_2 = \begin{bmatrix} 0 \\ 1 \\ 0 \end{bmatrix}; \quad \phi_3 = \begin{bmatrix} 0 \\ 0 \\ 1 \end{bmatrix} \tag{3.73}$$

as the basis.

Suppose now that the two subsystems interact in a way that gives rise to composite functions of the form

$$\chi = \sum_{i=1}^{4}\sum_{j=1}^{3} \alpha_{ij}(\psi_i \otimes \phi_j) \tag{3.74}$$

where operator $\otimes$ denotes the Kronecker product. Given a pair of vectors $a \in R^4$ and $b \in R^3$, this product can be defined as

$$a \otimes b = a \cdot b^T = \begin{bmatrix} a_1b_1 & a_1b_2 & a_1b_3 \\ a_2b_1 & a_2b_2 & a_2b_3 \\ a_3b_1 & a_3b_2 & a_3b_3 \\ a_4b_1 & a_4b_2 & a_4b_3 \end{bmatrix} \tag{3.75}$$

Although the Kronecker product is considerably simpler than the operator introduced in (3.67), there are a number of important similarities between them. Perhaps the most prominent one is that both operations produce aggregate states whose structure is considerably more complex than the structure of the subsystem states. This property is apparent from equation (3.74), which indicates that the states of the overall system must be described by 4×3 matrices, while those of S_1 and S_2 are adequately represented by vectors of dimension 4×1 and 3×1, respectively.

Another property is that is common to both operators is the "irreducibility" of the composite states. In our example, this is due to the fact that a 4×3 matrix χ can have rank as high as 3, while the rank of $x \cdot y^T$ is always equal to 1. As a result, χ can be expressed in the form

$$\chi = x \otimes y \tag{3.76}$$

only in very special cases. This clearly indicates that the overall system typically has properties that cannot be associated with the behavior of the subsystems. The "whole", in other words, is greater than the sum of its parts.

In addition to its explanatory role in the EPR paradox, the phenomenon of quantum entanglement also plays an important part in the so-called "measurement problem," which remains one of the most hotly debated issues in modern physics. A fundamental question that arises in this context is how measurement devices should be incorporated into the overall model. Some scientists believe that they should be viewed as quantum systems which interact with the particles whose properties they are designed to measure. Others have claimed that the behavior of the measurement apparatus is essentially classical, and that its description lies outside of quantum theory.

To see the relationship between quantum entanglement and measurements, suppose that we wish to determine the spin of an electron using an appropriate instrument. For the sake of argument, let us assume that the measurement device is a quantum system in its own right, with three eigenstates:

$$\begin{array}{lcl} \psi_0 & \leftrightarrow & \text{Instrument is idle} \\ \psi_1 & \leftrightarrow & \text{Instrument shows } \uparrow \\ \psi_2 & \leftrightarrow & \text{Instrument shows } \downarrow \end{array} \tag{3.77}$$

Since the electron can have two definite states of spin along the z - axis, we can describe it in terms of two eigenfunctions:

$$\begin{array}{lcl} \phi_1 & \leftrightarrow & \text{Spin-up} \\ \phi_2 & \leftrightarrow & \text{Spin-down} \end{array} \tag{3.78}$$

If we now decide to treat the measurement process as an interaction between the electron and the apparatus, what will be the outcome? In order to answer this question, let us first consider the simple case when the initial composite state of the system is

$$\chi_0 = \psi_0 \otimes \phi_1 \tag{3.79}$$

This corresponds to a situation where the device is idle prior to the interaction, and the electron is in a definite "spin-up" state. During the interaction, the wave function $\chi(t)$ evolves according to Schrödinger's equation, ultimately reaching the final state

$$\chi_f = \psi_1 \otimes \phi_1 \tag{3.80}$$

Such a state can obviously be related to the individual subsystems - the meter shows a definite reading $\uparrow$, which matches the definite spin-up state of the electron. This is very much what one would expect to see in a classical system.

Things become considerably more interesting, however, when the electron does *not* have a definite spin prior to the interaction. In that case, it is in a state of superposition, which can be described as

$$\phi = \alpha_1\phi_1 + \alpha_2\phi_2 \tag{3.81}$$

What will a measurement produce in this case? The state of the composite system will have the initial form

$$\chi_0 = \psi_0 \otimes (\alpha_1\phi_1 + \alpha_2\phi_2) \tag{3.82}$$

which can be meaningfully associated with subsystem states ψ_0 and $\phi = \alpha_1\phi_1 + \alpha_2\phi_2$, respectively. However, once the interaction is complete, the system will evolve into state

$$\chi_f = \alpha_1(\psi_1 \otimes \phi_1) + \alpha_2(\psi_2 \otimes \phi_2) \tag{3.83}$$

State χ_f is now *entangled*, and can *not* be meaningfully associated with the subsystem states. The electron and the measurement device are in a "joint" state of superposition, in which $|\alpha_1|^2$ and $|\alpha_2|^2$ represent the probabilities that we will record states $\chi_1 = \psi_1 \otimes \phi_1$ and $\chi_2 = \psi_2 \otimes \phi_2$, respectively, when the wave function χ_f collapses. Then (and only then) can we give an interpretation of the measurement that is consistent with our everyday experience.

3.4 Group Theory and Supersymmetry

The final section of this chapter is devoted to the notion of symmetry, which plays a vital role in quantum mechanics (as well as in particle physics and string theory). The principal tool for studying this property is *group theory*, whose main ideas will be outlined in the following. Although this outline is very limited and simplified, it should hopefully provide the reader with a sense of how symmetry is formally incorporated into modern physics. [5]

Symmetry Groups

To gain a basic understanding of what is meant by a *symmetry group*, consider the square shown in Fig. 3.3. What types of transformations will leave this object unchanged?

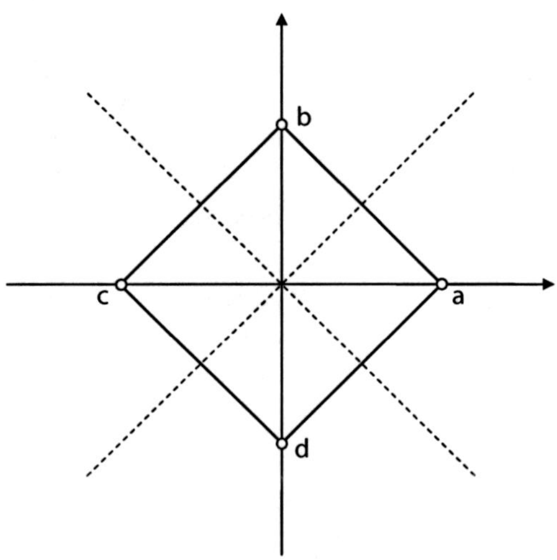

Fig. 3.3. The square and its axes of symmetry.

The square will obviously remain invariant to rotations around the origin by $0°, 90°, 180°$ and $270°$. It is also unchanged when reflected (or 'flipped') around the four axes indicated in Fig. 3.3.

A convenient way of representing these eight 'symmetry preserving' operations is to view vertices $\{a, b, c, d\}$ as points in the complex plain. The original configuration can then be represented as a complex vector

$$z = \begin{bmatrix} a \\ b \\ c \\ d \end{bmatrix} = \begin{bmatrix} 1 \\ i \\ -1 \\ -i \end{bmatrix} \tag{3.84}$$

It is not difficult to see that a rotation of the square by 90° is equivalent to multiplying z by the imaginary unit. Indeed, the new vector

$$w = iz = \begin{bmatrix} a \\ b \\ c \\ d \end{bmatrix} = \begin{bmatrix} i \\ -1 \\ -i \\ 1 \end{bmatrix} \tag{3.85}$$

corresponds to the reordering of vertices shown in Fig. 3.4. In a similar way, it can be shown that rotations by 180° and 270° correspond to $w = -z$ and $w = -iz$, respectively.

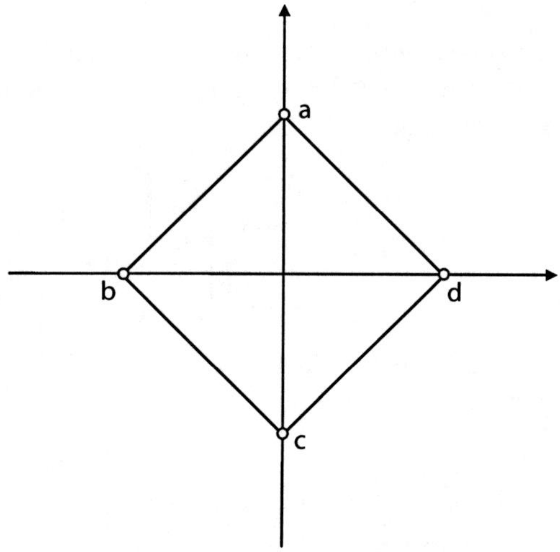

Fig. 3.4. The square after rotation.

Reflections around the horizontal and vertical axes can be represented in terms of the complex conjugate of z (in the following, we will use the symbol C to denote this operation). For the horizontal axis, we have

$$w = Cz = \begin{bmatrix} 1 \\ -i \\ -1 \\ i \end{bmatrix} \tag{3.86}$$

which produces the ordering of vertices shown in Fig. 3.5 (note that the operation $-Cz$ corresponds to a reflection around the vertical axis).

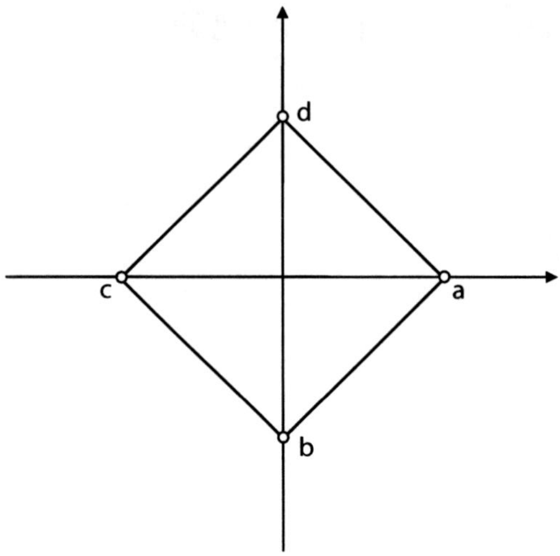

Fig. 3.5. The square after reflection around the horizontal axis.

Reflections around the remaining two axes are somewhat more complicated, and require a *composite* operation. This operation consists of multiplication by i, followed by complex conjugation:

$$w = C(iz) = C \begin{bmatrix} i \\ -1 \\ -i \\ 1 \end{bmatrix} = \begin{bmatrix} -i \\ -1 \\ i \\ 1 \end{bmatrix} \tag{3.87}$$

The resulting arrangement is shown in Fig. 3.6. Following the same procedure, it is easily verified that the operation

$$w = -C(iz) = \begin{bmatrix} i \\ 1 \\ -i \\ -1 \end{bmatrix} \tag{3.88}$$

corresponds to the reflection shown in Fig. 3.7.

The Formal Definition of a Group

The collection of eight transformations

$$G = \{g_1, \ldots, g_8\} = \{1, i, -1, -i, C, -C, Ci, -Ci\} \tag{3.89}$$

represents what is known as a *group*. Since these operation leave the original object unchanged, they are usually referred to as a *symmetry group*.

From a formal standpoint, a group can be viewed as a collection of mathematical operations with the following properties.

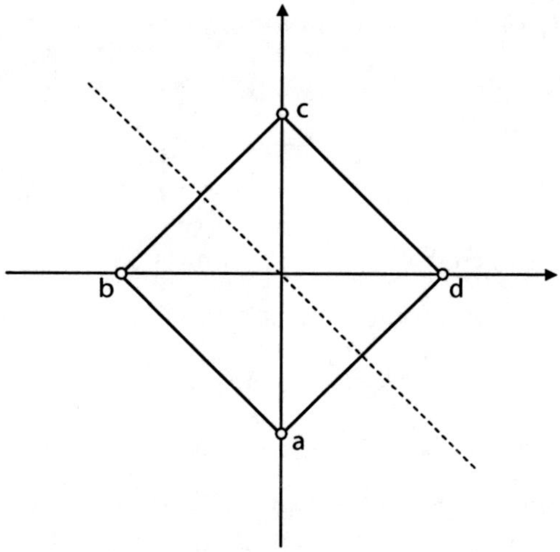

Fig. 3.6. Reflection around one of the diagonal axes.

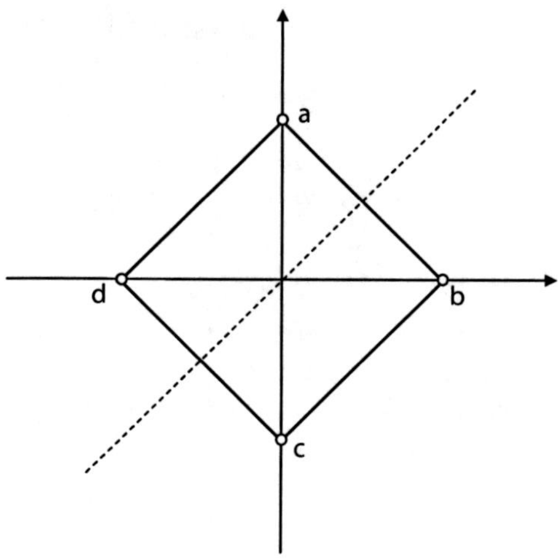

Fig. 3.7. Reflection around the other diagonal axis.

Property 1. The successive application of two group members, g_i and g_j, defines the operation of *multiplication* in the group. The resulting element $g = g_i g_j$ must also be a member of the group.

Example 3.6. Consider the successive application of $g_2 = i$ and $g_7 = Ci$, both of which belong to group G. The resulting operation is

$$w = g_2 g_7 z = g_2 \begin{bmatrix} -i \\ -1 \\ i \\ 1 \end{bmatrix} = \begin{bmatrix} 1 \\ -i \\ -1 \\ i \end{bmatrix} \tag{3.90}$$

It is easily verified that the same w can be obtained by applying $g_4 = C$

$$w = g_4 z = \begin{bmatrix} 1 \\ -i \\ -1 \\ i \end{bmatrix} \tag{3.91}$$

In other words, we have that $g_2 g_7 = g_4$, and $g_4 \in G$.

Property 2. The operation of multiplication is *associative*:

$$(g_i g_j) g_k = g_i (g_j g_k) \tag{3.92}$$

Property 3. The group must have an *identity element*, I, such that

$$gI = g \tag{3.93}$$

for all $g \in G$. In our group, $g_1 = 1$ is obviously such an element.

Property 4. Each element in the group must have an *inverse element*, $g_i^{-1} \in G$, which satisfies

$$g_i g_i^{-1} = I \tag{3.94}$$

As an illustration, consider element $g_4 = -i$ of group G. Since $g_4 i = 1$, it obviously follows that $g_4^{-1} = i = g_2$, which is also an element of the group.

Subgroups and the Abelian Property

Subgroups play a significant role in particle physics and unification theories. A subgroup is simply a subset of the original group, which constitutes a group in its own right. In the case of group G defined in (3.89), typical examples would be

$$S_1 = \{1, i, -1, -i\} \tag{3.95}$$

or

$$S_2 = \{1, -1, C, -C\} \tag{3.96}$$

It is easily verified that the elements of S_1 and S_2 satisfy Properties 1-4.

Subgroups often have special properties that the whole group does not. For example, not all elements of group G satisfy

$$g_i g_j = g_j g_i \tag{3.97}$$

Indeed, if we were to test this property on elements $g_5 = C$ and $g_2 = i$, we would obtain

$$g_5 g_2 = Ci = -iC = -g_2 g_5 \tag{3.98}$$

which is a violation of (3.97). However, if we focus on subgroup S_1 *only*, (3.97) *does* hold for all $g_i, g_j \in S_1$. We refer to subgroups with such a property as *Abelian groups* (after the Norwegian mathematician Niels Abel).

Representation of a Finite Group

When dealing with simple objects such as the square in Fig. 3.3, the symmetry preserving transformations can be represented by elementary mathematical operations such as multiplication by a complex number or conjugation. However, for more complicated and abstract objects, such a representation is usually inadequate. A possible way to describe a general finite group is to associate matrices $T(g_1), T(g_2), \dots T(g_n)$ with each of its members. For this to be a legitimate representation, the matrices have to preserve the multiplication law of the original group. In other words, for any two elements $g_i, g_j \in G$, we must have

$$T(g_i)T(g_j) = T(g_i g_j) \tag{3.99}$$

How can we construct matrices that conform to (3.99)? A simple way to do this is to label the rows and columns of $T(g_k)$ as

$$T(g_k) = \begin{array}{c} \\ g_1 \\ g_2 \\ \vdots \\ g_n \end{array} \begin{array}{c} \begin{matrix} g_1 & g_2 & \cdots & g_n \end{matrix} \\ \begin{bmatrix} t_{11} & t_{12} & \cdots & t_{1n} \\ t_{21} & t_{22} & \cdots & t_{2n} \\ \vdots & \vdots & \ddots & \vdots \\ t_{n1} & t_{n2} & \cdots & t_{nn} \end{bmatrix} \end{array} \tag{3.100}$$

and define its elements as

$$t_{ij} = \begin{cases} 1, & \text{if } \ g_i g_j^{-1} = g_k \\ 0, & \text{otherwise} \end{cases} \tag{3.101}$$

It is not difficult to verify that this construction secures the validity of (3.99).

Example 3.7. In the following, we will show that (3.99) holds for the representation of subgroup $S_1 = \{1, i, -1, -i\}$. In order to do this, let us (arbitrarily) pick $g_2 = i$ and $g_3 = -1$, and form matrices

$$T(g_2) = \begin{array}{c} \\ 1 \\ i \\ -1 \\ -i \end{array} \begin{array}{c} \begin{matrix} 1 & i & -1 & -i \end{matrix} \\ \begin{bmatrix} 0 & 0 & 0 & 1 \\ 1 & 0 & 0 & 0 \\ 0 & 1 & 0 & 0 \\ 0 & 0 & 1 & 0 \end{bmatrix} \end{array} \tag{3.102}$$

and

$$T(g_3) = \begin{array}{c} \\ 1 \\ i \\ -1 \\ -i \end{array} \begin{array}{c} \begin{matrix} 1 & i & -1 & -i \end{matrix} \\ \begin{bmatrix} 0 & 0 & 1 & 0 \\ 0 & 0 & 0 & 1 \\ 1 & 0 & 0 & 0 \\ 0 & 1 & 0 & 0 \end{bmatrix} \end{array} \tag{3.103}$$

Since $g_2 g_3 = -i = g_4$, and

$$T(g_4) = \begin{array}{c} \\ 1 \\ i \\ -1 \\ -i \end{array} \begin{array}{c} \begin{array}{cccc} 1 & i & -1 & -i \end{array} \\ \begin{bmatrix} 0 & 1 & 0 & 0 \\ 0 & 0 & 1 & 0 \\ 0 & 0 & 0 & 1 \\ 1 & 0 & 0 & 0 \end{bmatrix} \end{array} \tag{3.104}$$

it is easily verified that

$$T(g_2)T(g_3) = T(g_2 g_3) = T(g_4) \tag{3.105}$$

holds.

It is important to recognize that symmetry groups can be *finite* or *continuous.* Group G described in (3.89) is obviously finite, since it contains only eight elements. An example of a continuous group would be the rotation of a circle around its center. In this case, rotation by *any* angle α leaves the object unchanged. Furthermore, since successive rotations by angles α and β are equivalent to a single rotation by angle $\alpha + \beta$, it is not difficult to show that this set constitutes a group. Continuous groups are very common in quantum mechanics, particularly in quantum field theory. The mathematical framework that provides a description for such groups is known as *Lie algebra*, after the Norwegian mathematician Sophus Lie.

Supersymmetry

A proper examination of group theory and its role in modern physics clearly lies beyond the scope of this book, since it requires a much more advanced mathematical background than the one we have provided. Nevertheless, I feel it is instructive to close this chapter with an example which illustrates how group theory is used in quantum mechanics. Although our discussion will be brief and informal, it should nevertheless provide the reader with some important insights on this subject.

Our focus in the following will be on quantum field theory, which provides the necessary conceptual framework for particle physics. This theory makes extensive use of symmetry groups, which allow us (among other things) to mathematically transform one set of particles into another. One of the main difficulties that is encountered in this context stems from the fact that conventional symmetry groups (of the sort that we discussed) do *not* allow us to transform bosons into fermions, and vice versa. This restriction poses a major theoretical challenge, since it often leads to results that make no physical sense (such as infinite probabilities, for example).

A possible way to circumvent this problem is to modify the standard properties of a group, and expand the rules of Lie algebra so that they satisfy both anticommutation and commutation laws. Such extensions have led to generalized concepts such as *supergroups* and *supersymmetry*, which allow us to eliminate the troublesome infinities. This becomes possible once each fermion is assigned a bosonic *superpartner*[6] (and vice versa), whose presence largely neutralizes the underlying mathematical difficulties. The idea that all known particles have "superpartners" may perhaps sound strange, but it is widely accepted by physicists (although the existence of such entities has yet to be confirmed experimentally).[7]

Perhaps the most striking consequence of supersymmetry is the possibility that the electromagnetic, strong nuclear and weak nuclear forces could be indistinguishable at very high energies. Models that do not incorporate supersymmetry lack such unifying power, and predict instead that the strengths of the three forces would be similar, but not *identical.* This is clearly a prospect that most theoretical physicists find thoroughly unappealing (both aesthetically and philosophically). We should add in this context that aesthetic considerations are taken very seriously in science, since they are often our best criterion for evaluating hypotheses that cannot be verified experimentally. Unification theories certainly fall into that category, since the energy levels required to test them are impossible to reproduce (it is believed that such extreme physical conditions existed only in the immediate aftermath of the Big Bang).

3.5 Notes

1. For operators, eigenfunctions play the same role that eigenvectors play for matrices.

2. A scalar product is a rule by which two elements of a linear space produce a *number* (either real or complex). If these elements happen to be real vectors, then the scalar product can be defined as

$$\langle x, y\rangle = x^T y \tag{3.106}$$

If, on the other hand, the elements are complex-valued functions, we can use

$$\langle f(x), g(x)\rangle = \int f^*(\rho)g(\rho)d\rho \tag{3.107}$$

since this integral produces a complex number (note that $f^*(\rho)$ represents the complex conjugate of $f(\rho)$).

3. This property follows directly from the definition of the scalar product, since

$$\begin{aligned}&\langle \sum_i a_i f_i(x), g(x)\rangle = \textstyle\int (\sum_i a_i f_i(\rho))^* g(\rho)d\rho = \\ &= \sum_i a_i^* \textstyle\int f_i^*(\rho)g(\rho)d\rho = \sum_i a_i^* \langle f_i(x), g(x)\rangle\end{aligned} \tag{3.108}$$

4. Operators can have infinitely many distinct eigenvalues (unlike $n \times n$ matrices which have no more than n). In general, the set of eigenvalues that correspond to a linear operator can be *discrete* or *continuous.*

5. There are a number of excellent introductory texts on group theory. Some suggestions for further reading are provided below:

(a) Morton Hammermesh, *Group Theory and Its Applications to Physical Problems*, Dover Publications, 1990.

(b) Derek Robinson, *A Course in the Theory of Groups*, Springer-Verlag, 1996.

(c) Roy McWeeny, *Symmetry: An Introduction to Group Theory and Its Applications*, Dover Publications, 2002.

6. Some examples of superpartner pairs are: electron-*selectron*, photon-*photino*, and neutrino-*sneutrino*.

7. Scientists hope that the Large Hadron Collider (LHC) may be able to resolve some of these issues. As the world's largest particle accelerator, the LHC could potentially provide the energy levels needed to confirm the existence of superpartners.

Further Reading

L. D. Landau and L. M. Lifshitz, *Quantum Mechanics: Non-Relativistic Theory*, Butterworth-Heinemann, 1981.

John Wheeler and Wojciech Zurek, *Quantum Theory and Measurement*, Princeton University Press, 1984.

John Bell, *Speakable and Unspeakable in Quantum Mechanics*, Cambridge University Press, 2004.

David Bohm, *Quantum Theory*, Dover, 1989.

Chris Isham, *Quantum Theory: Mathematical and Structural Foundations*, Imperial College Press, 1995.

Werner Heisenberg, *Physics and Philosophy: The Revolution in Modern Science*, Harper Perennial, 2007.

Paul Dirac, *Lectures on Quantum Mechanics*, Dover, 2001.

Karl Popper, *Quantum Theory and the Schism in Physics*, Routledge, 1992.

R. Shankar, *Principles of Quantum Mechanics*, Springer, 1994.

John Polkinghorne, *Quantum Physics and Theology: An Unsuspected Kinship*, Yale University Press, 2008.

Henry Stapp, *Mindful Universe: Quantum Mechanics and the Participating Observer*, Springer, 2007.

Hugh Everett, *Many-Worlds Interpretation of Quantum Mechanics*, Princeton University Press, 1973.

Chapter 4

The Theory of Relativity

In a span of about 12 years, Albert Einstein produced two theories that brought about a fundamental change in our understanding of space, time, matter and energy. The results that he obtained are so profoundly counterintuitive that they cannot be fully grasped without the help of rather sophisticated mathematics. In the following sections, we will describe this formalism in some detail, and will consider a number of key results that follow from it. Before we do that, however, it may be helpful to briefly describe some of the most important implications of special and general relativity.

Einstein's theory of special relativity grew out of two remarkably simple postulates, which can be paraphrased as follows:

Postulate 1. The laws of physics must be identical for all observers moving at constant velocities.

Postulate 2. The speed of light is the same for all observers moving at constant velocities.

The first of these postulates is not particularly surprising, since we intuitively expect that the laws of nature ought to be independent of the coordinate system from which they are observed. The second one, however, appears to be at odds with our everyday experience, which suggests that objects moving toward us appear faster, while those moving away seem to be slower. There was no apparent reason to assume that light would be the lone exception to this rule. And yet, Einstein's bold conjecture turned out to be correct, and ultimately lead to some remarkable discoveries.

Special relativity makes extensive use of the so-called Lorentz transformation, which establishes a precise mathematical connection between measurements made by observers who are in relative motion. One of the key assumptions behind this transformation is that two observers will record *different times* and *different spatial locations* for the same event. This is quite unlike the paradigm used in classical physics, where time is identical for all observers (although the spatial coordinates that they register may vary).

A particularly interesting property of the Lorentz transformation is that the position x_A recorded by observer A generally depends on *both* the time *and* the position recorded by observer B (the same holds true for time t_A as well). The fact that temporal and spatial coordinates somehow become "entangled" as we move from one frame of reference to another has far-reaching consequences. Among other things, it suggests that we must forego our natural tendency to treat space and time as separate entities, and should instead view them as components of a unified four dimensional *spacetime*.

Although the Lorentz transformation is just a mathematical operation, it turns out that it has some very real physical manifestations. One of the best known examples of this kind is the so-called *Lorentz contraction*, which implies that a stationary observer will record a *shorter* length for an object than his moving counterpart. In the extreme case when the object moves at the speed of light relative to the observer, it becomes impossible to distinguish between its "front" and "back".

An even more striking illustration of the way in which Lorentz transformations alter our conventional ideas about time and space is an effect known as *time dilation*, which implies that the time interval recorded by a moving observer will be *shorter* than the one measured by his stationary counterpart. As the speed of the moving observer approaches the speed of light, his clock will actually trickle down to a halt (which is clearly something that is difficult to imagine without the help of mathematics).

Einstein's theory of gravity (which goes by the name of general relativity) poses a different sort of challenge to our intuition. One of its most important claims is that space is actually an *active* entity, and not a mere "container" through which material objects move. Einstein's conjecture was that spacetime and matter *interact*, and that this process can be described in precise mathematical terms. According to this hypothesis, spacetime becomes warped in the presence of matter (or energy), and its geometric form is determined by the way in which the masses are distributed. In such a framework, the notion of a gravitational "force" becomes obsolete, since the trajectory of a physical object is dictated by the curvature of spacetime.

The idea that space is curved is not all that difficult to grasp, since we happen to be living on the surface of a sphere (which is definitely not flat). Curved time, on the other hand, is much harder to imagine, since we have no similar experiences in our daily lives. In order to do this, we must first reconcile ourselves to the possibility that time can flow faster or slower, depending on the local geometric properties. Einstein's theory predicts, for example, that time will "slow down" for us if we happen to be in the vicinity of a massive object (where spacetime is significantly curved). As we move away from it, the curvature of spacetime becomes less pronounced, and our clocks will run faster.

Perhaps the most remarkable feature of general relativity is the fact that it provides a mathematical framework for studying the universe as a whole. It goes without saying, of course, that such a framework is necessarily incomplete, since there is no way to accurately describe a system of such complexity. Nevertheless, Einstein's equations allow us to formulate a meaningful approximate model of the universe, and to make certain predictions. It is interesting to note, however, that this model also points to the existence of truths which will necessarily remain unknown to us. A typical example of this sort is the recognition that some remote parts of the universe may have laws and constants of nature that are very different from the ones we are familiar with. Unfortunately (or perhaps fortunately), this is something that we will never be able to verify, since these regions are now permanently beyond our visible horizon.

To understand why this is a realistic possibility, we need to briefly describe a cosmological model proposed by physicist Alan Guth, which is commonly referred to as *inflation theory*. Guth suggested that the universe emerged from a miniscule "nugget" of spacetime which consisted of locally uniform "patches". He then found a solution of Einstein's equations that allows for a brief period of exponential expansion, during which each "patch" experienced a dramatic increase in size (this supposedly occurred in the immediate aftermath of the Big Bang). According to inflation theory, the part of the universe that we now inhabit most probably originated from a tiny segment of the primordial "nugget" which was small enough for particles to interact and homogenize. Such a prediction is certainly consistent with our astronomical data, which suggest that the "observable" universe is remarkably uniform in every respect. It is entirely possible, however, that the other "patches" had very different physical properties, and that the regions which grew out of them are nothing like our own. Given that no information from these regions will ever reach us (due to the expansion of the universe), the only thing that we can say with any reliability is that is that our theories and mathematical models apply *locally*.

Another intriguing consequence of Einstein's general relativity is the existence of black holes. Since the entire mass of these unusual heavenly bodies is crushed to the size of a point, the curvature of spacetime at their center approaches infinity (this extreme scenario is commonly referred as a *singularity*). Within a certain region around the singularity, the pull of gravity is so strong that nothing can escape, not even light. This property clearly places a limit on what we can know about such objects, since it becomes impossible to acquire empirical information about processes that take place in their interior. It is interesting to note, however, that our ignorance on this subject is actually a very good thing, since the same "shielding effect" that prevents us from fully understanding black holes also prohibits them from influencing the outside world. If it were not for this restriction, black holes would directly affect the area around them, and our universe would be a very different (and undoubtedly far less hospitable) place.

In the following sections, we will examine some of these questions in greater detail. We will first consider special relativity, and summarize some of its most important results. We will then turn our attention to general relativity, and describe how gravity can be analyzed in the framework of differential geometry.

4.1 Special Relativity

Special relativity is concerned with the way in which physical processes appear to different observers. This problem dates back (at least) to Galileo, who was the first to mathematically connect observations made in reference systems that are in relative motion. The simplest way to introduce Galilean transformations is to consider a moving object which is observed from two different coordinate systems, A and B, as shown in Fig. 4.1. One of these systems is assumed to be stationary, while the other moves with a *constant* velocity $\vec{V}_R$.

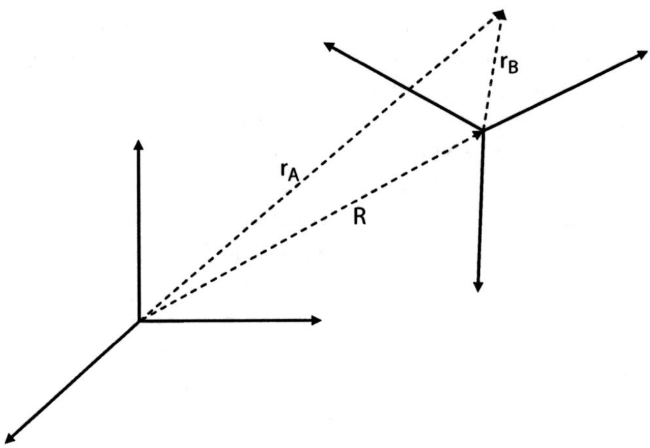

Fig. 4.1. Two coordinate systems in relative motion.

Since

$$\vec{r}_A(t) = \vec{r}_B(t) + \vec{R}(t) \tag{4.1}$$

and $\vec{R}(t) = \vec{V}_R\, t$, we have

$$\vec{r}_A(t) = \vec{r}_B(t) + \vec{V}_R\, t \tag{4.2}$$

In the one dimensional case, this simplifies to

$$x_A(t) = x_B(t) + v_R t \tag{4.3}$$

It is important to recognize that from the Galilean perspective time is *identical* in the two coordinate systems (that is, $t_A = t_B = t$). Differentiating both sides of (4.3) over time, we obtain

$$v_A(t) = v_B(t) + v_R \tag{4.4}$$

which leads to the conclusion that the velocity of the object will generally be *different* for the two observers. However, since

$$a_A(t) = \frac{dv_A}{dt} = \frac{dv_B}{dt} = a_B(t) \tag{4.5}$$

the acceleration will remain the same.

One of the most important characteristics of any coordinate transformation is quantities that remain *invariant* to it. It is not difficult to show that for Galilean transformations the *distance* between two objects exhibits this property. Indeed, if two coordinate systems move with a relative velocity v_R, the distance between objects P and Q can be expressed as

$$x_A^{(p)} - x_A^{(q)} = x_B^{(p)} + v_R t - \left[x_B^{(q)} + v_R t\right] = x_B^{(p)} - x_B^{(q)} \tag{4.6}$$

The preservation of distance allows us to introduce a common *metric* for all frames of reference. In three dimensions, this metric can be formulated in terms of the displacement between two nearby points as

$$ds^2 = dx^2 + dy^2 + dz^2 \tag{4.7}$$

The Lorentz Transformation

Although the Galilean approach worked very well in the context of Newtonian mechanics, in the late 1800s it was observed that Maxwell's equations are *not* invariant to such transformations. This posed a conceptual problem, since basic electromagnetic phenomena ought to be unaffected by the relative motion of the observers. Einstein stated this principle explicitly in the form of two postulates:

Postulate 1. The laws of physics must be identical for all observers moving at constant velocities.

Postulate 2. The speed of light is the same for all observers moving at constant velocities.

In order for these two postulates to hold, it was necessary to introduce a new coordinate transformation under which Maxwell's equations would remain invariant. It turns out that this requirement is satisfied by the so-called *Lorentz transformation*, which is based on the assumption that two observers will record *different times* and *different spatial locations* for the same event. This is quite unlike the Galilean view, where it was assumed that all observers record the *same* time for an event (the difference being only in the spatial coordinates).

In describing the Lorentz transformation, we will consider two observers whose relative speed is v_R, and whose record of the same event is given in terms of spacetime coordinates (t_A, x_A) and (t_B, x_B), respectively (for simplicity, we will consider only one spatial dimension). In matrix form, the relationship between these coordinates is given as

$$\begin{bmatrix} ct_B \\ x_B \end{bmatrix} = L \begin{bmatrix} ct_A \\ x_A \end{bmatrix} \tag{4.8}$$

where

$$L = \Gamma \begin{bmatrix} 1 & -v_R/c \\ -v_R/c & 1 \end{bmatrix} \tag{4.9}$$

and

$$\Gamma = \frac{1}{\sqrt{1 - v_R^2/c^2}} \tag{4.10}$$

We can rewrite (4.8) and its inverse in a more explicit form as

$$\begin{aligned} t_B &= \Gamma\left[t_A - (v_R/c^2)x_A\right] \qquad & t_A &= \Gamma\left[t_B + (v_R/c^2)x_B\right] \\ x_B &= \Gamma\left[-v_R t_A + x_A\right] \qquad & x_A &= \Gamma\left[v_R t_B + x_B\right] \end{aligned} \tag{4.11}$$

Note that in the limiting case when $v_R << c$ (which is typical for Newtonian mechanics) we have $\Gamma \approx 1$, which means that (4.11) reduces to Galilean transformations. This is of fundamental importance, since it preserves the validity of results established by classical mechanics for low velocities.

The Lorentz transformation has an invariance property which has far reaching consequences. To see this, let us consider two events that are described by observer A in terms of spacetime coordinates (t_A, x_A) and $(t_A + dt_A, x_A + dx_A)$. In the framework of observer B, the same two events have coordinates (t_B, x_B) and $(t_B + dt_B, x_B + dx_B)$. From (4.8) we directly obtain

$$\begin{bmatrix} cdt_B \\ dx_B \end{bmatrix} = L \begin{bmatrix} cdt_A \\ dx_A \end{bmatrix} \tag{4.12}$$

Observing that

$$c^2dt_B^2 - dx_B^2 = \begin{bmatrix} cdt_B & dx_B \end{bmatrix} \begin{bmatrix} 1 & 0 \\ 0 & -1 \end{bmatrix} \begin{bmatrix} cdt_B \\ dx_B \end{bmatrix} \tag{4.13}$$

and that

$$L^T \begin{bmatrix} 1 & 0 \\ 0 & -1 \end{bmatrix} L = \begin{bmatrix} 1 & 0 \\ 0 & -1 \end{bmatrix} \tag{4.14}$$

we now have

$$c^2dt_B^2 - dx_B^2 = c^2dt_A^2 - dx_A^2 \equiv ds^2 \tag{4.15}$$

The quantity ds is known as the *interval*, and is invariant under Lorentz transformations.

Equation (4.15) is of fundamental importance for special relativity. It points to the fact that ds^2 represents a kind of distance measure (or metric) in spacetime which is *identical* in all reference frames. The four dimensional extension of (4.15)

$$ds^2 = c^2dt^2 - dx^2 - dy^2 - dz^2 \tag{4.16}$$

represents a metric that defines the so-called *Minkowski spacetime*. In such a space, the vector $[ct\ x\ y\ z]^T$ provides a natural representation for any given event (the superscript T denotes the transpose of a matrix).

It is important to recognize that the interval ds depends on *local* coordinates only, and is independent of the relative motion of the two reference frames. This allows us to introduce the notion of *proper time* as

$$d\tau = ds/c \tag{4.17}$$

Proper time defined in this way is a quantity that is common to *all* observers, since it can be computed directly from (4.16) as

$$d\tau^2 = dt^2 - \frac{1}{c^2}\left(dx^2 + dy^2 + dz^2\right) \tag{4.18}$$

in an arbitrary frame of reference. This definition also allows us to express the Minkowski metric in a slightly more convenient way, as

$$c^2d\tau^2 = c^2dt^2 - dx^2 - dy^2 - dz^2 \tag{4.19}$$

It should be noted that the spacetime metric defined in (4.16) is rather unusual, since it allows for the possibility that $ds^2 < 0$ (which would correspond to something like a "complex distance"). In a typical three dimensional Euclidean space this is not possible, since the metric has the form

$$ds^2 = dx^2 + dy^2 + dz^2 \tag{4.20}$$

where all signs are *positive*. The space described in (4.20) is said to have *signature* $(+,+,+)$, which is quite unlike the Minkowski space whose signature is $(+,-,-,-)$. This is an important property, which defines the scalar product of vectors $x = [x_1\ x_2\ x_3\ x_4]^T$ and $y = [y_1\ y_2\ y_3\ y_4]^T$ in spacetime as

$$x \cdot y = x_1 y_1 - x_2 y_2 - x_3 y_3 - x_4 y_4 \tag{4.21}$$

The Minkowski spacetime is said to be *flat*, since the metric (4.16) has constant coefficients $\left(c^2, -1, -1, -1\right)$ next to the differential terms. In an n-dimensional *curved* space, ds^2 would have a more general form

$$ds^2 = \sum_{i,j} g_{ij}(x) dx_i dx_j \tag{4.22}$$

where the coefficients in the metric vary from point to point (there is also the possibility of cross product terms $dx_i dx_j$). One of the immediate consequences of the "flatness" of Minkowski spacetime is the fact that the shortest distance between two points is a straight line. This is not the case with curved spaces, as will become apparent in Section 4.2.

The following two examples illustrate the profound way in which Lorentz transformations alter our conventional ideas about time and space.

Example 4.1. (The Lorentz Contraction). Let us assume that an extended one dimensional object (a ruler, for example) is moving with velocity v_R with respect to a stationary observer A. Another observer, B, is assumed to move alongside the object under consideration. In order to measure the length of the ruler, let us fix a light source at each of its endpoints, and assume that these sources send out a simultaneous flash. For observer B (who sees the object as motionless) these two events will obviously occur at the *same* time, and their spatial coordinates will be $x_B^{(1)}$ and $x_B^{(2)}$. Since $t_B^{(1)} = t_B^{(2)}$, he can directly compute the length as

$$l_0 = x_B^{(2)} - x_B^{(1)} \tag{4.23}$$

It is important to recognize, however, that observer A will *not* see the two flashes simultaneously, since for him the object is moving. Instead, he will record these events at times $t_A^{(1)}$ and $t_A^{(2)}$, respectively. To emphasize this point, in the following we will denote the the corresponding spatial coordinates of these events by $x_A^{(1)}(t_A^{(1)})$ and $x_A^{(2)}(t_A^{(2)})$.

According to the Lorentz transformation (with $t_B^{(1)} = t_B^{(2)}$ included), the measurements of the two observers are related as

$$\Delta t_A = t_A^{(2)} - t_A^{(1)} = \Gamma(v_R/c^2)\left(x_B^{(2)} - x_B^{(1)}\right) = \Gamma(v_R/c^2) l_0 \tag{4.24}$$

and

$$x_A^{(2)}(t_A^{(2)}) - x_A^{(1)}(t_A^{(1)}) = \Gamma\left(x_B^{(2)} - x_B^{(1)}\right) = \Gamma l_0 \tag{4.25}$$

Recalling that $x_A^{(1)}(t_A^{(1)})$ and $x_A^{(2)}(t_A^{(2)})$ are recorded at *different* points in time, it follows that observer A cannot determine the length of the object by simply subtracting $x_A^{(1)}(t_A^{(1)})$ from $x_A^{(2)}(t_A^{(2)})$. For this measurement to be meaningful, he needs to compute l as

$$l = x_A^{(2)}(t_A^{(2)}) - x_A^{(1)}(t_A^{(2)}) \tag{4.26}$$

which means that he must know the values for $x_A^{(1)}$ and $x_A^{(2)}$ at the *same* point in time. The easiest way to recover this information is to add and subtract $x_A^{(1)}(t_A^{(2)})$ from the left hand side of (4.25), which produces

$$\left[x_A^{(2)}(t_A^{(2)}) - x_A^{(1)}(t_A^{(2)})\right] + \left[x_A^{(1)}(t_A^{(2)}) - x_A^{(1)}(t_A^{(1)})\right] = \Gamma l_0 \tag{4.27}$$

Since each endpoint moves uniformly with velocity v_R, the difference $x_A^{(1)}(t_A^{(2)}) - x_A^{(1)}(t_A^{(1)})$ can be computed as

$$x_A^{(1)}(t_A^{(2)}) - x_A^{(1)}(t_A^{(1)}) = v_R\left(t_A^{(2)} - t_A^{(1)}\right) \tag{4.28}$$

which allows us to rewrite (4.27) as

$$l + v_R \Delta t_A = \Gamma l_0 \tag{4.29}$$

Using the expression for Δt_A in given (4.24), we finally obtain

$$l = \Gamma l_0 - v_R \Delta t_A = \Gamma l_0 - \Gamma(v_R^2/c^2) l_0 = l_0\sqrt{1 - v_R^2/c^2} \tag{4.30}$$

According to (4.30), the stationary observer A will record a shorter length than his moving counterpart. This effect, usually referred to as the *Lorentz contraction*, suggests that geometric shapes are "distorted" in some sense when the observed object is moving. We should note that the contraction is negligible at velocities that are much smaller than the speed of light, since in that case $v_R/c \approx 0$. However, as $v_R \to c$, we have that $l \to 0$. In other words, to an observer moving at the speed of light relative to the object, the distinction between its "front" and "back" vanishes.

Example 4.2. (Time Dilation). Let us assume that observer A is stationary, while observer B is travelling in a space ship which moves with velocity v_R. Both observers record the time when the ship departs, and when it arrives at its destination. What will their clocks show? According to the Lorentz transformation,

$$\begin{aligned} t_B^{(2)} &= \Gamma\left[t_A^{(2)} - (v_R/c^2)x_A^{(2)}\right] \\ t_B^{(1)} &= \Gamma\left[t_A^{(1)} - (v_R/c^2)x_A^{(1)}\right] \end{aligned} \tag{4.31}$$

and therefore

$$t_B^{(2)} - t_B^{(1)} = \Gamma \left[t_A^{(2)} - t_A^{(1)} - (v_R/c^2) \left(x_A^{(2)} - x_A^{(1)} \right) \right] \tag{4.32}$$

Since the spaceship travels with a constant velocity v_R, observer A can compute the overall distance to the destination as

$$x_A^{(2)} - x_A^{(1)} = v_R \left(t_A^{(2)} - t_A^{(1)} \right) \tag{4.33}$$

Substituting this into (4.32) and recalling the definition of Γ, we directly obtain

$$\Delta t_B = t_B^{(2)} - t_B^{(1)} = \sqrt{1 - v_R^2/c^2} \left(t_A^{(2)} - t_A^{(1)} \right) \tag{4.34}$$

It follows that the elapsed time recorded by the moving observer B is shorter than the time measured by the stationary one. In the extreme case when $v_R \to c$, we have that $\Delta t_B \to 0$, which means that time effectively stops. Thus, for an observer moving at the speed of light, the past, present and future are indistinguishable.

Inertial Observers

Before proceeding to discuss further implications of special relativity, we need to precisely define what is meant by the term *inertial observer*. All results presented up to this point referred to observers that move with *constant* velocities. As Einstein postulated, for these observers the laws of physics must be identical. But what about those whose motion is accelerated - do they experience physical processes in a different way? To gain an intuitive understanding of the issues involved, imagine that you are moving with a constant velocity in a sealed compartment. Under such circumstances, there is no way for you to determine what your speed is, or even whether you are moving at all. If, on the other hand, you happen to be accelerating, you will feel a definite force (such as the one that "pins" you against the seat of the car, or the centrifugal force in the case of rotation). These forces are real, but are not experienced by observers whose motion is uniform. They are often referred to as *kinematic* forces, in order to emphasize the fact that they originate from acceleration. Thus, if you happen to be moving with acceleration $\vec{a}$, the kinematic force that you will experience is:

$$\vec{F}_K = -m\vec{a} \tag{4.35}$$

The minus sign indicates that the force has a direction that is *opposite* to the actual acceleration.

As a further illustration of this phenomenon, let us consider an observer who is located in a freely falling container somewhere near the surface of the earth. Since she is moving downward with an acceleration equal to $\vec{g}$, she will experience a kinematic force $\vec{F}_K = -m\vec{g}$. Note, however, that this force actually cancels out the gravitational force $\vec{F}_g = m\vec{g}$, so the measurements of this observer will show *zero acceleration*. What Einstein realized was that *all* observers whose accelerometer measures zero are equivalent, provided that existing kinematic forces are taken into account. Such observers are referred to as *inertial.* The fact that physical phenomena seen by an accelerating observer are indistinguishable from those witnessed by one who moves with a constant velocity is quite remarkable, and represents one of the fundamental symmetries of nature. The following example underscores the special role that gravity plays in this context.

Example 4.3. To better illustrate the difference between inertial and non-inertial observers, let us imagine a moving individual who is holding an accelerometer in his hand. For the sake of simplicity, we will assume that the meter consists of a spring and a mass m_1 that is suspended from it. In that case, any extension or compression of the spring corresponds to a nonzero reading of the instrument.

Let us now assume that our observer is moving in a *non-gravitational* field, in which the force acting on any object depends on the object's position (but *not* on its mass). Since the observer and the accelerometer have the same spatial location $\vec{r}$ (measured with respect to some stationary reference point), they will experience the *same* force, $\vec{F}(\vec{r})$. Note, however, that if the mass of the observer (m_0) and the meter (m_1) are *not* the same, the two accelerations

$$\vec{a}_0(\vec{r}) = \frac{\vec{F}(\vec{r})}{m_0} \tag{4.36}$$

and

$$\vec{a}_1(\vec{r}) = \frac{\vec{F}(\vec{r})}{m_1} \tag{4.37}$$

will be *different.* As a result, the observer will record two unequal forces that act on mass m_1: $\vec{F}(\vec{r})$, and the *kinematic* force

$$\vec{F}_K = -m_1\vec{a}_0(\vec{r}) = -\frac{m_1}{m_0}\vec{F}(\vec{r}). \tag{4.38}$$

Since these forces do not cancel each other out, the spring must stretch in order to achieve an equilibrium, and the observer will record a nonzero reading on his accelerometer. Based on this reading, he will conclude that his system of reference is *not* inertial.

Let us now contrast this situation to the one where the observer is freely falling in a *gravitational* field, whose source is some object with a very large mass M. Although the observer and the meter are located at the same distance r from this mass, they will experience different forces

$$\vec{F}_0 = \gamma\frac{m_0 M}{r^2} \tag{4.39}$$

and

$$\vec{F}_1 = \gamma\frac{m_1 M}{r^2} \tag{4.40}$$

Their accelerations, however, will be *exactly* the same, since

$$\vec{a}_0(\vec{r}) = \frac{\vec{F}_0}{m_0} = \gamma\frac{M}{r^2} = \frac{\vec{F}_1}{m_1} = \vec{a}_1(\vec{r}) \tag{4.41}$$

This property ensures that the gravitational force acting on the meter, $\vec{F}_1$, will be completely cancelled out by its kinematic counterpart

$$\vec{F}_K = -m_1\vec{a}_0(\vec{r}) = -\gamma\frac{m_1 M}{r^2} \tag{4.42}$$

In this case, the observer will record no extension of the spring, and will conclude that his system of reference is *inertial.*[1]

Relativistic Mechanics

The Lorentz transformation was introduced to ensure the invariance of Maxwell's equations under coordinate changes. However, according to Einstein's postulate, *all* laws of physics must have this property. This implies that the laws of Newtonian mechanics need to be modified, since their classical formulation is *not* invariant under Lorentz transformations. In making the appropriate adjustments, we will also require that the new "relativistic" mechanics must reduce to Newton's laws at low velocities. This is clearly necessary, since classical mechanics provides an accurate representation of reality under such conditions.

As a first step, it is useful to recall the basic equations of classical mechanics as they relate to the concepts of force, momentum and energy. The standard formulation of Newton's second law has the form

$$\vec{F} = \frac{d\vec{p}}{dt} = m_0\frac{d\vec{v}}{dt} \tag{4.43}$$

Rewriting this as

$$m_0 d\vec{v} = \vec{F}dt \tag{4.44}$$

and performing a scalar multiplication of both sides by vector $\vec{v}$, we obtain

$$\frac{1}{2}m_0 d(v_x^2 + v_y^2 + v_z^2) = (F_x v_x + F_y v_y + F_z v_z)dt \tag{4.45}$$

Observing that the expression on the right hand side represents the incremental increase in total energy, we have

$$d\left(\frac{m_0 v^2}{2}\right) = dE \tag{4.46}$$

This equation is easily integrated to obtain

$$E = E_0 + \frac{m_0 v^2}{2} \tag{4.47}$$

where E_0 is an arbitrary constant. In classical mechanics we typically set $E_0 = 0$, so that the total energy of a free particle equals its kinetic energy. If we were to allow $E_0 \neq 0$, this would imply that a motionless particle (with no external field present) has some sort of "permanent" potential energy.

In order to preserve invariance with respect to Lorentz transformations, it is necessary to reformulate Newton's second law in four dimensional spacetime as

$$\frac{d\vec{\Pi}}{d\tau} = \vec{\Phi} \tag{4.48}$$

where $\vec{\Pi}$ and $\vec{\Phi}$ represent generalized notions of momentum and force, respectively, while τ denotes proper time. As in classical mechanics, the generalized momentum can be expressed as

$$\vec{\Pi} = m_0 \vec{U} \tag{4.49}$$

where

$$\vec{U} = \begin{bmatrix} u_0 \\ u_1 \\ u_2 \\ u_3 \end{bmatrix} = \frac{d}{d\tau} \begin{bmatrix} ct(\tau) \\ x(\tau) \\ y(\tau) \\ z(\tau) \end{bmatrix} \tag{4.50}$$

represents the generalized velocity. This definition of $\vec{U}$ makes sense when we recall that any event in spacetime is uniquely determined by the vector $[ct\ x\ y\ z]^T$. We should also note in this context that it is natural to describe the evolution of spacetime trajectories in terms of proper time τ, since this quantity is common to all observers.

Recalling the Minkowski metric (4.19) and dividing both sides by dt^2, we obtain

$$c^2 \left(\frac{d\tau}{dt}\right)^2 = c^2 - v_x^2 - v_y^2 - v_z^2 = c^2 - v^2 \tag{4.51}$$

and therefore

$$\frac{dt}{d\tau} = \frac{1}{\sqrt{1 - v^2/c^2}} \equiv \gamma \tag{4.52}$$

This allows us to rewrite (4.50) in a more convenient form as

$$\vec{U} = \frac{d}{dt} \begin{bmatrix} ct \\ x \\ y \\ z \end{bmatrix} \cdot \frac{dt}{d\tau} = \begin{bmatrix} \gamma c \\ \gamma v_x \\ \gamma v_y \\ \gamma v_z \end{bmatrix} \equiv \begin{bmatrix} u_0 \\ u_1 \\ u_2 \\ u_3 \end{bmatrix} \tag{4.53}$$

Observing that the scalar product in Minkowski spacetime

$$\vec{U} \cdot \vec{U} = u_0^2 - u_1^2 - u_2^2 - u_3^2 = \gamma^2[c^2 - v^2] = c^2 \tag{4.54}$$

is *constant*, it follows that

$$0 = \frac{d}{d\tau}\left(\vec{U} \cdot \vec{U}\right) = 2\left[u_0 \frac{du_0}{d\tau} - u_1 \frac{du_1}{d\tau} - u_2 \frac{du_2}{d\tau} - u_3 \frac{du_3}{d\tau}\right] \tag{4.55}$$

If we denote the components of the generalized force $\vec{\Phi}$ by $[\phi_0\ \phi_1\ \phi_2\ \phi_3]^T$, equation (4.48) allows us to rewrite (4.55) as

$$\frac{u_0\phi_0}{m_0} - \frac{u_1\phi_1}{m_0} - \frac{u_2\phi_2}{m_0} - \frac{u_3\phi_3}{m_0} = 0 \tag{4.56}$$

By (4.48), we also have that

$$\phi_1 = m_0\frac{du_1}{d\tau} = \frac{d}{d\tau}(m_0\gamma v_x) = \frac{d}{dt}(m_0\gamma v_x)\frac{dt}{d\tau} = \gamma\frac{d}{dt}(m_0\gamma v_x) \tag{4.57}$$

In the following, we will introduce the notation

$$F_x = \frac{d}{dt}(m_0\gamma v_x) \tag{4.58}$$

which allows us to express ϕ_1 in compact form as

$$\phi_1 = \gamma F_x \tag{4.59}$$

The new notation is logical, since for $\gamma \to 1$ (4.58) becomes a restatement of Newton's second law, and F_x turns into a classical force. The same reasoning applies to the other two components, ϕ_2 and ϕ_3, with

$$F_y = \frac{d}{dt}(m_0\gamma v_y); \qquad F_z = \frac{d}{dt}(m_0\gamma v_z) \tag{4.60}$$

Given the proposed interpretation of ϕ_1, ϕ_2 and ϕ_3, it remains to identify an appropriate expression for component ϕ_0. We can do this by rewriting (4.56) as

$$u_0\phi_0 = c\gamma\phi_0 = u_1\phi_1 + u_2\phi_2 + u_3\phi_3 \tag{4.61}$$

and recognizing that

$$\begin{aligned} u_1\phi_1 &= (\gamma v_x)(\gamma F_x) \\ u_2\phi_2 &= (\gamma v_y)(\gamma F_y) \\ u_3\phi_3 &= (\gamma v_z)(\gamma F_z) \end{aligned} \tag{4.62}$$

Component ϕ_0 can now be expressed as

$$\phi_0 = \frac{\gamma}{c}(F_x v_x + F_y v_y + F_z v_z) \tag{4.63}$$

On the other hand, from (4.48) it follows that

$$\phi_0 = m_0\frac{du_0}{d\tau} = \frac{d}{d\tau}(m_0 c\gamma) = \gamma\frac{d}{dt}(m_0 c\gamma) \tag{4.64}$$

Equating (4.63) and (4.64), we obtain

$$(F_x v_x + F_y v_y + F_z v_z)\, dt = d(m_0 c^2\gamma) \tag{4.65}$$

The left hand side of this expression can be viewed as generalized incremental energy which reduces to classical energy when $\gamma \to 1$. We can therefore write

$$dE = d(m_0 c^2\gamma) \tag{4.66}$$

and integrate to obtain Einstein's famous relation

$$E = mc^2 \tag{4.67}$$

where the mass of the particle

$$m = m_0\gamma = \frac{m_0}{\sqrt{1 - v^2/c^2}} \tag{4.68}$$

depends on its velocity. In interpreting this result, it is important to recognize that there is a constant term $E_0 = m_0 c^2$ even when the particle is *not* moving. This is quite unlike classical mechanics, where the energy of a free particle at rest is zero. In the relativistic model there is an *intrinsic* potential energy that is directly related to the mass of the particle, which allows us to view matter and energy as equivalent and interchangeable quantities. Equation (4.68) also points to the fact that no material object can move with the speed of light. Indeed, if we simultaneously had $m_0 \neq 0$ and $v \to c$, this would result in $m \to \infty$, which is physically impossible.

Remark 4.1. In (4.67) the integration constant was set to zero, which seems like a rather "artificial" choice. Couldn't we have picked this constant so that $E_0 = 0$, as in the case in classical mechanics? The answer to this question is no, although the reasoning behind it is rather subtle. For our purposes, it suffices to note that the Lorentz transformation must reduce to its Galilean counterpart when $\gamma \to 1$. It turns out that this requirement would be violated if a nonzero integration constant was included in (4.67).

4.2 General Relativity

In the following, we will briefly consider some basic concepts and results that are related to general relativity. Since the proper mathematical framework for studying this subject is *differential geometry*, we begin with an outline of some of its main ideas.

Curved Spaces

As a first step toward understanding Einstein's theory, it is important to recognize that the "medium" for gravitational interaction is not a conventional force, but rather *spacetime itself.* With that in mind, it is perhaps easier to grasp the intimate relationship that exists between gravity and the geometry of curved spaces.

As is usually the case in geometry, it makes sense to start with the notion of *distance.* The main question that we need to consider in this context is how distances can be expressed in terms of curvilinear coordinates. In standard n-dimensional Euclidean coordinates, the displacement between two neighboring points can be expressed as a vector

$$ds = \begin{bmatrix} ds_1 \\ \vdots \\ ds_n \end{bmatrix} \tag{4.69}$$

The corresponding metric (which is basically a measure of distance) can be defined as

$$ds^2 = ds^T ds = ds_1^2 + \ldots + ds_n^2 \tag{4.70}$$

If we wish to express this displacement in terms of a set of *curvilinear* coordinates $x_1, x_2, \ldots, x_n$, it is convenient to rewrite ds_i as

$$ds_i = \sum_{k=1}^{n} \frac{\partial s_i}{\partial x_k} dx_k \tag{4.71}$$

In that case, we obtain

$$ds = H(x)dx \tag{4.72}$$

where

$$H(x) = \begin{bmatrix} \partial s_1/\partial x_1 & \cdots & \partial s_1/\partial x_n \\ \vdots & \ddots & \vdots \\ \partial s_n/\partial x_1 & \cdots & \partial s_n/\partial x_n \end{bmatrix}; \qquad dx = \begin{bmatrix} dx_1 \\ \vdots \\ dx_n \end{bmatrix} \tag{4.73}$$

Denoting the elements of matrix

$$G(x) = H^T(x)H(x) \tag{4.74}$$

by $g_{ij}(x)$, the metric can be expressed in terms of the curvilinear coordinates as

$$ds^2 = dx^T H^T(x)H(x)dx = \sum_{i,\, j} g_{ij}(x)dx_i dx_j \tag{4.75}$$

Note that in this case the differential terms are *not* multiplied by constants, but rather by functions of x. There is also the possibility of cross-product terms $dx_i dx_j$ $(i \neq j)$.

Example 4.4. To illustrate the transition from Euclidean to curvilinear metrics, let us consider the connection between Cartesian and spherical coordinates, which is given as

$$\begin{aligned} x &= r \sin\theta \cos\phi \\ y &= r \sin\theta \sin\phi \\ z &= r \cos\theta \end{aligned} \tag{4.76}$$

The standard Euclidean metric is obviously

$$ds^2 = dx^2 + dy^2 + dz^2 \tag{4.77}$$

To obtain the corresponding curvilinear representation, we can differentiate (4.76) with respect to θ, ϕ and r, obtaining

$$H(x) = \begin{bmatrix} r\cos\theta\cos\phi & -r\sin\theta\sin\phi & \sin\theta\cos\phi \\ r\cos\theta\sin\phi & r\sin\theta\cos\phi & \sin\theta\sin\phi \\ -r\sin\theta & 0 & \cos\theta \end{bmatrix} \tag{4.78}$$

It is not difficult to verify that

$$G(x) = H^T(x)H(x) = \begin{bmatrix} r^2 & 0 & 0 \\ 0 & r^2 \sin^2\theta & 0 \\ 0 & 0 & 1 \end{bmatrix} \tag{4.79}$$

Since this turns out to be a diagonal matrix, there will be no cross product terms, and the metric can be expressed as

$$ds^2 = r^2 d\theta^2 + r^2 \sin^2\theta d\phi^2 + dr^2 \tag{4.80}$$

Shortest Distance in Curved Spaces

Given a pair of points A and B in a curved space, it is often of interest to determine the shortest distance between them. In flat spaces, this distance obviously corresponds to a straight line. In the general case of metric (4.75), however, the answer is by no means straightforward. To consider this issue in more detail, let us first note that any curve in an n-dimensional space can be represented *parametrically* as

$$x(\sigma) = \begin{bmatrix} x_1(\sigma) \\ \vdots \\ x_n(\sigma) \end{bmatrix} \tag{4.81}$$

For any value of the scalar parameter σ, we can thus determine the corresponding point on the curve. How should we pick σ? In the context of dynamic systems the role of σ is played by time t, and a trajectory is defined by specifying $x_1(t), \ldots, x_n(t)$. This representation subsequently allows us to plot the curve in the state space, as discussed in Chapter 1. In the more general scenario that we are currently dealing with, σ can be pretty much anything. One possibility that will prove particularly useful is to choose it as the *length* of the curve measured from A to the present point.

In order to determine whether a particular curve represents the shortest distance between points A and B (which correspond to parameter values σ_1 and σ_2, respectively), we need to compare it to other curves that pass through these same points. To that effect, we will consider a perturbation of $x(\sigma)$ in the form $x(\sigma) + \varepsilon\rho(\sigma)$, where $\rho(\sigma) = [\rho_1(\sigma) \ldots \rho_n(\sigma)]^T$ is an arbitrary function and ε is a small number. If $x(\sigma)$ is indeed the shortest path, then we must have

$$\left.\frac{dS}{d\varepsilon}\right|_{\varepsilon=0} = 0 \tag{4.82}$$

where S is the length of this curve calculated from A to B. In other words, we require that the curve length is *minimal* for $\varepsilon = 0$ (which corresponds to the unperturbed $x(\sigma)$).

To rewrite condition (4.82) in a more convenient way, we must first consider how the total length S is computed. One way to do this is to rewrite the metric as

$$ds^2 = \sum_{i,j} g_{ij}(x) dx_i dx_j = \sum_{i,j} g_{ij}(x) \dot{x}_i \dot{x}_j d\sigma^2 \tag{4.83}$$

where

$$\dot{x}_i = \frac{dx_i}{d\sigma} \tag{4.84}$$

Introducing a scalar *Lagrangian*

$$L(x, \dot{x}) = \sum_{i,j} g_{ij}(x)\dot{x}_i\dot{x}_j \tag{4.85}$$

an infinitesimal distance in the curved space can be expressed as

$$ds = \sqrt{L(x, \dot{x})}d\sigma \tag{4.86}$$

This allows us to determine the overall length S by integrating along the curve

$$S = \int_{\sigma_1}^{\sigma_2} \sqrt{L(x, \dot{x})}d\sigma \tag{4.87}$$

For minimization purposes, we need

$$\frac{dS}{d\varepsilon} = \int_{\sigma_1}^{\sigma_2} \frac{d}{d\varepsilon}\left(\sqrt{L(x, \dot{x})}\right) d\sigma = 0 \tag{4.88}$$

to hold for $\varepsilon = 0$. After some algebra, it turns out that

$$\left.\frac{dS}{d\varepsilon}\right|_{\varepsilon=0} = \frac{1}{2}\sum_{i=1}^{n}\int_{\sigma_1}^{\sigma_2}\left\{\frac{d}{d\sigma}\left[-\frac{1}{\sqrt{L}}\frac{\partial L}{\partial \dot{x}_i}\right] + \frac{1}{\sqrt{L}}\frac{\partial L}{\partial x_i}\right\}\rho_i(\sigma)d\sigma \tag{4.89}$$

Since this integral must be zero for *any* choice of $\rho(\sigma)$, it follows that $x(\sigma)$ is the shortest path between A and B if this curve satisfies

$$\frac{d}{d\sigma}\left[\frac{1}{\sqrt{L}}\frac{\partial L}{\partial \dot{x}_i}\right] - \frac{1}{\sqrt{L}}\frac{\partial L}{\partial x_i} = 0 \qquad i = 1, 2, \ldots, n \tag{4.90}$$

Curves for which condition (4.90) holds are known as *geodesics*. Since such curves represent the shortest path between two points, they are in many ways analogous to straight lines in flat spaces. With that in mind, it is reasonable to expect that in a curved space light will move along a geodesic (which is precisely what Einstein's general relativity claims).

Example 4.5. As noted earlier, one of the simplest curved spaces is the surface of a unit sphere. It is therefore instructive to consider what geodesics might look like in such a space. The position of any point on the surface of a sphere can be described by two angles, θ and ϕ (since the radius r is fixed and equal to 1). It is well known that the corresponding metric has the form (see Example 4.4)

$$ds^2 = \sin^2\theta d\phi^2 + d\theta^2 \tag{4.91}$$

from which we can easily identify $g_{\phi\phi} = \sin^2\theta$ and $g_{\theta\theta} = 1$ (the off-diagonal coefficients $g_{\phi\theta}$ and $g_{\theta\phi}$ are zero).

An obvious solution for (4.90) is to equate the Lagrangian L to a *constant.* In the special case when $L = 1$, it follows that

$$S = \int_{\sigma_1}^{\sigma_2} \sqrt{L(x,\dot{x})}d\sigma = \int_{\sigma_1}^{\sigma_2} d\sigma \tag{4.92}$$

so parameter σ can be interpreted as the length along the curve (measured from the initial point that corresponds to σ_1). Since

$$L(x,\dot{x}) = \sum_{i,j} g_{ij}(x)\dot{x}_i\dot{x}_j = \left(\sin^2\theta\right)\dot{\phi}^2 + \dot{\theta}^2 \tag{4.93}$$

one of the conditions for the geodesic is

$$\left(\sin^2\theta\right)\dot{\phi}^2 + \dot{\theta}^2 = 1 \tag{4.94}$$

A second condition follows from the fact that L *does not* depend explicitly on ϕ. As a result

$$\frac{\partial L}{\partial \phi} = 0 \tag{4.95}$$

and therefore (by virtue of (4.90))

$$\frac{d}{d\sigma}\left[\frac{1}{\sqrt{L}}\frac{\partial L}{\partial \dot{\phi}}\right] = 0 \tag{4.96}$$

This implies that

$$\frac{\partial L}{\partial \dot{\phi}} = 2\dot{\phi}\sin^2\theta = J \tag{4.97}$$

where J is an unspecified constant.

For a given initial condition (θ_0, ϕ_0), equations (4.94) and (4.97) can be solved explicitly in terms of σ, producing

$$\begin{aligned} \cos\theta(\sigma) &= \cos\theta_0 \sin(\sigma - \sigma_0) \\ \tan(\phi(\sigma) - \phi_0) &= \sin\theta_0 \tan(\sigma) \end{aligned} \tag{4.98}$$

Although it is not easy to visualize this, (4.98) actually describes a *great circle* that passes through points A and B. Thus, geodesics in this space correspond to circles whose center and radius match those of the unit sphere itself.

Parallel Transport

Before we proceed to discuss Einstein's equations, we must also briefly consider the notion of *parallel transport* in curved spaces. In a flat space we can take any pair of vectors $\vec{a}$ and $\vec{b}$ and "slide" them along a curve so that they remain parallel to their originals, as illustrated in Fig. 4.2. It is important to recognize

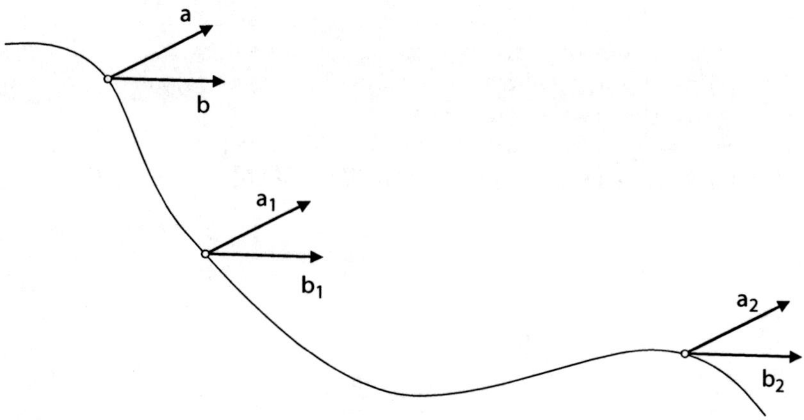

Fig. 4.2. Parallel transport in flat space.

that in this process, the scalar product $\vec{a} \cdot \vec{b}$ is preserved, as is the length of both vectors.

It turns out that this process cannot be replicated in a curved space. The standard notion of parallelism does not work in this case, since the transported vector must lie in the space at all times. To get a better sense of the underlying problem, consider the surface of a unit sphere and a tangent on one of its great circles. The tangent is associated with point P_1 in Fig. 4.3, which represents an "aerial" view of the great circle. We can legitimately claim that vector $\vec{V}_1$ belongs to the curved surface, since it lies in one of the tangent planes. If we now parallel-slide the vector along the great circle to point P_2, the new vector $\vec{V}_2$ is *not* in the tangent plane any more. As such, it doesn't belong to the surface of the sphere, and cannot represent the parallel transport of vector $\vec{V}_1$ in this space.

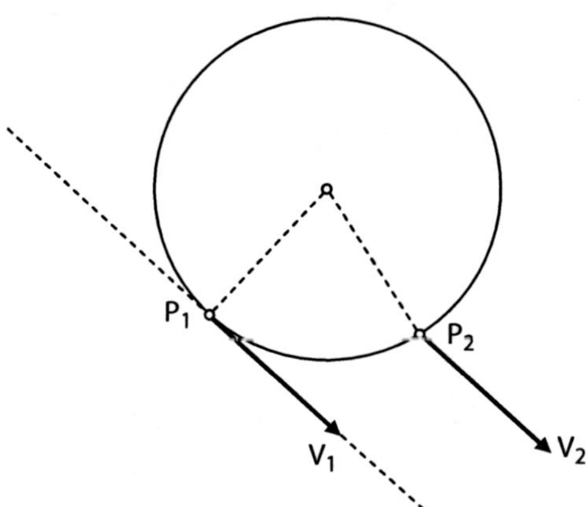

Fig. 4.3. Parallel-sliding on the surface of a sphere.

In view of the difficulties noted above, it is necessary to provide a different notion of parallel transport that is suitable for curved spaces. In doing so, we will still require that the scalar product and lengths of vectors $\vec{a}$ and $\vec{b}$ remain preserved, as was the case in flat spaces.

Before proposing a new definition for parallel transport, it is useful to reformulate the Lagrange equations (4.90). Recalling (4.85) it is not difficult to show that

$$\frac{\partial L}{\partial \dot{x}_\mu} = 2\sum_\nu g_{\mu\nu}(x)\dot{x}_\nu \tag{4.99}$$

and

$$\frac{\partial L}{\partial x_\mu} = \sum_{\alpha,\,\beta} \frac{\partial g_{\alpha\beta}}{\partial x_\mu}(x)\dot{x}_\alpha \dot{x}_\beta \tag{4.100}$$

respectively. Substituting these expressions into (4.90), we obtain (after some rather tedious algebra) a compact reformulation of Lagrange's equations in the form

$$\ddot{x}_\mu + \sum_{\alpha,\,\beta} \Gamma^\mu_{\alpha\beta}\dot{x}_\alpha \dot{x}_\beta = 0 \qquad \mu = 1, 2, \ldots, n \tag{4.101}$$

The coefficients

$$\Gamma^\mu_{\alpha\beta} = \frac{1}{2}\sum_\nu g^{\mu\nu}(x)\left[\frac{\partial g_{\nu\alpha}}{\partial x_\beta} + \frac{\partial g_{\nu\beta}}{\partial x_\alpha} - \frac{\partial g_{\alpha\beta}}{\partial x_\nu}\right] \tag{4.102}$$

are known as *Christoffel symbols*, which play an important part in the formulation of the basic equations of general relativity (note that the terms $g^{\mu\nu}$ with superscript indices represent elements of matrix $G^{-1}(x)$, which are different from $g_{\mu\nu}$).

Example 4.6. Christoffel symbols are obviously difficult to compute directly from the definition. However, since they represent the multipliers next to $\dot{x}_\alpha$ and $\dot{x}_\beta$ in the reformulated Lagrange equations (4.101), it is not too hard to obtain them indirectly. To see what that means in practical terms, let us consider the surface of a sphere of radius R. As noted earlier, the metric of such a space has the form

$$ds^2 = R^2 \sin^2\theta d\phi^2 + R^2 d\theta^2 \tag{4.103}$$

with $g_{\phi\phi} = R^2\sin^2\theta$ and $g_{\theta\theta} = R^2$. The Lagrangian can then be computed as

$$L(x,\dot{x}) = \sum_{i,\,j} g_{ij}(x)\dot{x}_i\dot{x}_j = R^2\left(\sin^2\theta\right)\dot{\phi}^2 + R^2\dot{\theta}^2 \tag{4.104}$$

In their original form (4.90), Lagrange's equations for this space are

$$\frac{d}{d\sigma}\left[\frac{\partial L}{\partial\dot{\theta}}\right] - \frac{\partial L}{\partial\theta} = \frac{d}{d\sigma}\left(2R^2\dot{\theta}\right) - 2R^2\left(\sin\theta\cos\theta\right)\dot{\phi}^2 = 0 \tag{4.105}$$

and

$$\frac{d}{d\sigma}\left[\frac{\partial L}{\partial\dot{\phi}}\right] - \frac{\partial L}{\partial\phi} = \frac{d}{d\sigma}\left[2R^2\left(\sin^2\theta\right)\dot{\phi}\right] = 0 \tag{4.106}$$

Differentiating the first terms in (4.105) and (4.106) explicitly, we obtain

$$\begin{aligned} &\ddot{\theta} - (\sin\theta\cos\theta)\,\dot{\phi}^2 = 0 \\ &\ddot{\phi} + 2\,(\cot\theta)\,\dot{\theta}\dot{\phi} = 0 \end{aligned} \tag{4.107}$$

We can now find the Christoffel symbols by matching (4.107) to the generic form (4.101) and identifying the terms next to $\dot{x}_\alpha \dot{x}_\beta$.

The first equation has only the term $\dot{\phi}^2$, implying that

$$\Gamma^{\theta}_{\phi\phi} = -\sin\theta\cos\theta \tag{4.108}$$

while all other coefficients are zero. Similarly, from the second equation we obtain

$$\Gamma^{\phi}_{\theta\phi} = \Gamma^{\phi}_{\phi\theta} = \cot\theta \tag{4.109}$$

(the term $2\cot\theta$ was split in half to ensure the symmetry of matrix Γ^{ϕ}). All the remaining terms are zero.

We can now formally define parallel transport in curved spaces. Given a vector

$$V_A = \begin{bmatrix} v_A^{(1)} \\ \vdots \\ v_A^{(n)} \end{bmatrix} \tag{4.110}$$

and a curve

$$x(\sigma) = \begin{bmatrix} x_1(\sigma) \\ \vdots \\ x_n(\sigma) \end{bmatrix} \tag{4.111}$$

we will say that vector V_B represents the parallel transport of V_A to some neighboring point $x(\sigma + d\sigma) = x(\sigma) + \delta x$ if its components satisfy

$$v_B^{(\mu)} = v_A^{(\mu)} - \sum_{\alpha,\,\beta} \Gamma^{\mu}_{\alpha\beta} v_A^{(\beta)} \delta x_\alpha \qquad \mu = 1, 2, \ldots, n \tag{4.112}$$

Observing that $v_B^{(\mu)} - v_A^{(\mu)} \equiv \delta v_\mu$, and dividing both sides by $d\sigma$, this condition can be rewritten as a differential equation

$$\dot{v}_\mu + \sum_{\alpha,\,\beta} \Gamma^{\mu}_{\alpha\beta} \dot{x}_\alpha v_\beta = 0 \qquad \mu = 1, 2, \ldots, n \tag{4.113}$$

It is important to recognize that the *tangent* of $x(\sigma)$ satisfies this equation whenever the curve is a geodesic. Indeed, setting $v_\mu = \dot{x}_\mu$, we directly obtain

$$\ddot{x}_\mu + \sum_{\alpha,\,\beta} \Gamma^{\mu}_{\alpha\beta} \dot{x}_\alpha \dot{x}_\beta = 0 \qquad \mu = 1, 2, \ldots, n \tag{4.114}$$

which is automatically satisfied by virtue of (4.101).

Einstein's Equations in Empty Space

In order to formulate Einstein's equations in empty space, it is useful to first examine the corresponding equations that arise in Newtonian gravity. To that effect, let us consider a satellite that orbits the earth at a fixed distance. An observer located at the center of the satellite is *inertial*, since at this point the gravitational and centrifugal forces cancel out exactly. However, in points of the satellite that are radially removed from the center by some λ, the net acceleration will *not* be zero. This effect is due to the fact that the gravitational field is not homogeneous, and is not perfectly cancelled out the centrifugal force in all points. These residual forces are commonly referred to as *tidal forces.*

The components of the residual acceleration due to tidal forces can be expressed as

$$a_i(\lambda) = \sum_{j=1}^{3} \frac{\partial a_i}{\partial x_j} \lambda_j \tag{4.115}$$

Overall, the acceleration vector has the form

$$a(\lambda) = T\lambda \tag{4.116}$$

where matrix T is known as the *tidal tensor*. Recalling that in a gravitational field with potential φ the acceleration is given as

$$a_i = -\frac{\partial \varphi}{\partial x_i} \qquad i = 1, 2, 3 \tag{4.117}$$

we can express the elements of this tensor as

$$t_{ij} = \frac{\partial a_i}{\partial x_j} = -\frac{\partial^2 \varphi}{\partial x_i \partial x_j} \tag{4.118}$$

Newton's gravitation theory also requires that

$$\frac{\partial^2 \varphi}{\partial x_1^2} + \frac{\partial^2 \varphi}{\partial x_2^2} + \frac{\partial^2 \varphi}{\partial x_3^2} = 0 \tag{4.119}$$

which is equivalent to saying that the *trace* of the tidal tensor must equal zero (the trace of a matrix is the sum of its diagonal elements).

The equations of general relativity bear a close conceptual resemblance to the classical Newtonian equations (4.116) and (4.119). They are based on the recognition that an observer who travels along a geodesic trajectory in curved space is *inertial* (the fact that such an observer experiences no acceleration follows from Lagrange's equations (4.101), although not in an obvious way).

In order to formally state Einstein's equations, let us consider a curve which is obtained by parallel-transporting an infinitesimal vector λ along a geodesic. The "tip" of this vector traces out a curve that is not itself a geodesic. As a result, any objects that move along this curve will experience an acceleration that is given as

$$a_\mu(\lambda) = -\sum_{\alpha, \beta, \nu} R^{\mu}_{\ \nu\alpha\beta} \dot{x}_\alpha \dot{x}_\nu \lambda_\beta \equiv -\sum_{\beta} \Delta^{\mu}_{\beta} \lambda_\beta \tag{4.120}$$

The derivation of this expression is fairly complicated, and exceeds the scope of this book. For our purposes, it suffices to note that coefficients $R^{\mu}_{\nu\alpha\beta}$ represent elements of the so called *Riemann tensor.* Given a metric for spacetime, these elements can be computed from the Christoffel symbols.

A straightforward comparison reveals a clear resemblance between (4.120) and its classical counterpart (4.115). This similarity suggests that the terms

$$\Delta^{\mu}_{\beta} = \sum_{\alpha,\,\nu} R^{\mu}_{\nu\alpha\beta}\dot{x}_{\alpha}\dot{x}_{\nu} \tag{4.121}$$

can be viewed as elements of a *generalized tidal tensor.* By analogy with (4.119), the trace of this tensor should equal zero. In explicit form, this condition can be stated as

$$\sum_{\mu}\Delta^{\mu}_{\mu} = \sum_{\alpha,\,\beta}\left[\sum_{\mu} R^{\mu}_{\alpha\mu\beta}\right]\dot{x}_{\alpha}\dot{x}_{\beta} = 0 \tag{4.122}$$

Recalling that (4.122) must hold for all velocities $\dot{x}_i$, this condition reduces to

$$R_{\alpha\beta} = 0 \tag{4.123}$$

where

$$R_{\alpha\beta} = \sum_{\mu} R^{\mu}_{\alpha\mu\beta} \tag{4.124}$$

represent components of the so-called *Ricci tensor*.

The Schwarzschild Metric

Finding a metric that will result in a zero Ricci tensor is one of the central problems of general relativity. One of the few existing explicit solutions was proposed by Karl Schwarzschild in 1916. What makes this solution a remarkable tribute to the human spirit is the fact that it was discovered in the trenches of World War I. Schwarzschild was able to communicate his result to Einstein just months before his untimely death.

To understand the nature of Schwarzschild's result, let us first recall that the *flat* Minkowski metric of special relativity had the form

$$c^2d\tau^2 = c^2dt^2 - dx^2 - dy^2 - dz^2 \tag{4.125}$$

This expression can be rewritten in terms of spheric coordinates, using the relations

$$\begin{aligned} x &= r\sin\theta\cos\phi \\ y &= r\sin\theta\sin\phi \\ z &= r\cos\theta \end{aligned} \tag{4.126}$$

The new metric becomes

$$c^2d\tau^2 = c^2dt^2 - dr^2 - r^2d\theta^2 - r^2\sin^2\theta d\phi^2 \tag{4.127}$$

It is interesting to note that the change to spheric coordinates produces coefficients $g_{ij}(x)$ which are *not* constant, although the space itself is flat.

How would we expect this metric to change if spacetime is curved? Schwarzschild assumed the form

$$c^2 d\tau^2 = A(r)dt^2 - B(r)dr^2 - r^2 d\theta^2 - r^2 \sin^2 \theta d\phi^2 \tag{4.128}$$

and looked for coefficients $A(r)$ and $B(r)$ such that:

1. All elements of the Ricci tensor are zero.

2. The new metric must reduce to the flat Minkowski metric when $r \to \infty$. This requirement is equivalent to

$$\lim_{r\to\infty} A(r) = c^2 \qquad \lim_{r\to\infty} B(r) = 1 \tag{4.129}$$

Following a rather tedious (but well defined) procedure, Schwarzschild found that

$$A(r) = c^2(1 - a/r) \tag{4.130}$$

$$B(r) = \frac{1}{1 - a/r} \tag{4.131}$$

The parameter a in (4.130) and (4.131) is an unspecified integration constant that arises in the derivation. It is worth noting that this constant is intimately related to the *event horizon* of a black hole, which is the region around the center of the hole from which nothing can escape (see Example 4.9 for more details).

Schwarzschild's result is of great importance, and has produced some remarkable predictions. The following examples illustrate this in some detail.

Example 4.7. (Planetary Orbits). Since planetary orbits are assumed to be located in a single plane, we can set $\theta = \pi/2$ in (4.128) without loss of generality. This simplifies Schwarzschild's metric to the form

$$c^2 d\tau^2 = c^2(1 - a/r)dt^2 - \frac{dr^2}{1 - a/r} - r^2 d\phi^2 \tag{4.132}$$

from which we directly obtain

$$\begin{aligned} g_{tt} &= c^2(1 - a/r) \\ g_{rr} &= -(1 - a/r)^{-1} \\ g_{\phi\phi} &= -r^2 \end{aligned} \tag{4.133}$$

and therefore

$$L(x, \dot{x}) = c^2(1 - a/r)\dot{t}^2 - \frac{1}{1 - a/r}\dot{r}^2 - r^2\dot{\phi}^2 \tag{4.134}$$

as well. Since dividing both sides of (4.132) by $d\tau^2$ produces the same right hand side as in (4.134), it also follows that

$$L(x, \dot{x}) = c^2 \tag{4.135}$$

(note that proper time τ plays the role of parameter σ in this case).

Observing that L does not depend explicitly on t or ϕ, we have

$$\frac{d}{d\tau}\left[\frac{\partial L}{\partial \dot{t}}\right] = 0 \tag{4.136}$$

and

$$\frac{d}{d\tau}\left[\frac{\partial L}{\partial \dot{\phi}}\right] = 0 \tag{4.137}$$

These two conditions can be expressed explicitly as

$$(1 - a/r)\dot{t} = \gamma \tag{4.138}$$

$$r^2\dot{\phi} = J \tag{4.139}$$

where γ and J are unspecified constants. As a fourth equation, we can use

$$\dot{r} = \frac{dr}{d\tau} = \frac{dr}{d\phi}\dot{\phi} \tag{4.140}$$

Eliminating $\dot{t}$, $\dot{r}$ and $\dot{\phi}$ from (4.135),(4.138),(4.139) and (4.140), we obtain

$$\frac{J^2}{r^4}\left(\frac{dr}{d\phi}\right)^2 = c^2(\gamma^2 - 1) + \frac{ac^2}{r} - \frac{J^2}{r^2} + \frac{aJ^2}{r^3} \tag{4.141}$$

The solution of this equation becomes considerably simpler if we introduce a new variable $u = 1/r$. Such a modification allows us to rewrite (4.141) as

$$\left(\frac{du}{d\phi}\right)^2 = au^3 - u^2 + \alpha u + \beta \tag{4.142}$$

where

$$\alpha = \frac{ac^2}{J^2} \qquad \beta = \frac{c^2(\gamma^2 - 1)}{J^2} \tag{4.143}$$

This equation is easily separable as

$$\int_{\phi_0}^{\phi} d\phi = \int_{u_0}^{u} \frac{du}{(au^3 - u^2 + \alpha u + \beta)^{1/2}} \tag{4.144}$$

where u_0 and ϕ_0 represent a pair of points on the orbit.

An approximate analytic solution can be obtained by introducing quantities

$$r_1 = \frac{1}{u_1} \qquad r_2 = \frac{1}{u_2} \tag{4.145}$$

which correspond to points where

$$\frac{du}{d\phi} = 0 \tag{4.146}$$

Recalling (4.142), it follows that u_1 and u_2 are the two positive roots of polynomial

$$au^3 - u^2 + \alpha u + \beta \tag{4.147}$$

The approximate solution for the angle traversed between two successive maxima of u can now be expressed as

$$\phi - \phi_0 = 2\pi + \frac{3}{2}\pi a(u_1 + u_2) + O(a^2) \tag{4.148}$$

where $O(a^2)$ denotes higher order terms in a.

How can we interpret the result in (4.148)? In Newtonian mechanics, the trajectory of a planet is an *ellipse*, in which the orbiting object returns to point u_2 after traversing and angle 2π. In the case of general relativity, however, the angle traversed between to successive maxima of u *exceeds* 2π. It is as if the entire ellipse shifts by an angle equal to $\Delta\phi = 3\pi a(u_1 + u_2)/2$ during one revolution. Such an orbit is *not re-entrant*, and has actually been observed for the planet Mercury (where this anomaly is particularly noticeable). The angular deviation $\Delta\phi$ predicted by general relativity perfectly matches astronomical observations in this case.

Example 4.8. (Deflection of Light) The derivation in Example 4.7 ought to be valid for all particles in empty space, including photons. In this case, however, we must be careful with the way we choose constants γ, J and subsequently α and β as well. To begin with, we know that Schwarzschild's metric must reduce to

$$c^2 d\tau^2 = c^2 dt^2 - dx^2 - dy^2 - dz^2 \tag{4.149}$$

when $r \to \infty$. Dividing both sides of (4.149) by $d\tau^2$, we obtain the well known relationship of special relativity

$$\dot{t}^2 = \frac{1}{1 - v^2/c^2} \tag{4.150}$$

In the case of light $v = c$, so it follows that $\dot{t} \to \infty$. This has to hold for the general Schwarzschild metric as well, since it must produce the same result when $r \to \infty$.

One of the immediate implications of this fact is that $\gamma \to \infty$, since by (4.138)

$$\gamma = (1 - a/r)\dot{t} \tag{4.151}$$

Furthermore, since α and β must be finite in order to get a meaningful solution, we must require $J \to \infty$. In that case, we ensure that

$$\beta = \frac{c^2(\gamma^2 - 1)}{J^2} < \infty \tag{4.152}$$

and also

$$\alpha = \frac{ac^2}{J^2} \to 0 \tag{4.153}$$

With these modifications, equation (4.142) simplifies to

$$\left(\frac{du}{d\phi}\right)^2 = au^3 - u^2 + \beta \tag{4.154}$$

The polynomial on the right hand side now has only one positive root, u_1, and the integration

$$\int_{\phi_0}^{\phi} d\phi = \int_{u_0}^{u} \frac{du}{(au^3 - u^2 + \beta)^{1/2}} \tag{4.155}$$

produces an approximate solution

$$\phi - \phi_0 \approx \pi + 2au_1 \tag{4.156}$$

The interpretation of the result in (4.156) is that a beam of light will *deflect* as it passes by the sun (or any other massive object). It is interesting to note that this is precisely how Einstein's theory was validated. In 1919, British physicist Arthur Eddington observed a total eclipse of the sun during which stars were briefly visible in the daytime. According to general relativity, beams of light originating from the stars should be deflected by an angle of $\Delta\phi = 2au_1$ as they pass near the sun. The position of these stars would therefore appear to be different at night, when the sun's mass does not deform spacetime along the path of the light beams. The difference in the observed positions of the stars turned out to be exactly what Einstein had anticipated.

Black Holes

One of the most fascinating predictions of general relativity is the existence of black holes. These strange objects arise when a star collapses, and its entire mass is crushed to the size of a point. From the perspective of general relativity, this point represents a *singularity*, where time and space as we know them effectively cease to exist. The singularity is surrounded by an invisible boundary which is commonly referred to as the *event horizon.*

It is interesting to consider what an observer would experience if (by some miracle) he happened to reach the event horizon in one piece. For the sake of argument, suppose that the observer came from a space ship that is anchored at a safe distance from the black hole. Those watching from the space ship would actually never see this person reach the event horizon. For them, the whole process would slow down to a trickle, and would require an infinitely long time. The clock of the observer, on the other hand, would indicate nothing unusual, and would continue ticking as the event horizon is reached (and ultimately crossed). What the unfortunate observer would most definitely notice is a tremendous gravitational pull, due to the sharply changing curvature of spacetime. Since the curvature near the center of a black hole changes dramatically from point to point, the force pulling at his head would be much smaller than the one acting on his feet. As a result the observer's body would be stretched, and ultimately torn apart (at which point this turns into one of the more "gruesome" thought experiments of modern physics).

The following example illustrates some aspects of this strange phenomenon.

Example 4.9. For the sake of simplicity, let us assume that our observer is moving *radially* toward the center of the black hole (this means that coordinates θ and ϕ are fixed, and only r changes). In that case, the Schwarzschild metric reduces to

$$c^2 d\tau^2 = c^2(1-a/r)dt^2 - \frac{dr^2}{1-a/r} \tag{4.157}$$

and the Lagrangian has the form

$$L = c^2(1-a/r)\dot{t}^2 - \frac{\dot{r}^2}{1-a/r} = c^2 \tag{4.158}$$

Since L does not depend explicitly on t, condition (4.90) implies

$$\frac{d}{d\tau}\left[\frac{\partial L}{\partial \dot{t}}\right] = 0 \tag{4.159}$$

It is not difficult to verify that equation (4.159) is satisfied when

$$\dot{t} = \frac{\gamma}{1-a/r} \tag{4.160}$$

where γ is an arbitrary constant.

Setting $\gamma = 1$ (for simplicity) and substituting (4.160) into (4.158), we obtain the following differential equation

$$\dot{r}^2 = \frac{ac^2}{r} \tag{4.161}$$

In order to solve (4.160) and (4.161), it is convenient to introduce a new variable x, which is related to r as

$$r = ax^2 \tag{4.162}$$

In that case, (4.161) can be rewritten

$$\dot{r} = \frac{c}{x} \tag{4.163}$$

and differentiating (4.162) yields

$$\dot{r} = \frac{dr}{d\tau} = \frac{dr}{dx}\cdot\frac{dx}{d\tau} = 2ax\frac{dx}{d\tau} \tag{4.164}$$

Equating (4.163) and (4.164), we directly obtain

$$2ax^2 dx = c d\tau \tag{4.165}$$

Observing further that

$$1 - a/r = \frac{x^2-1}{x^2} \tag{4.166}$$

we can now rewrite (4.160) as

$$dt = \frac{d\tau}{1 - a/r} = \frac{2ax^4}{c(x^2 - 1)} dx \tag{4.167}$$

Integrating both sides of this expression, we finally obtain an explicit expression for t in terms of x

$$t = \frac{a}{c} \left(\frac{2}{3} x^3 + 2x + \log \frac{x - 1}{x + 1} \right) + B \tag{4.168}$$

where B is an arbitrary constant.

It is important to note at this point that the function $t(x)$ contains a *singularity* regardless of how B is chosen. This is due to the fact that $t \to \infty$ when $x \to -1$. Recalling the definition of x given in (4.162), it follows that this situation occurs when $r \to a$. Since t is the time recorded by observers in the space ship (and $t \to \infty$ as $r \to a$) it follows that they will never really see the other individual cross the imaginary boundary defined by $r = a$. This boundary represents the *event horizon* of the black hole.

Integrating (4.165), we can also express the proper time τ as [2]

$$\tau = \frac{2a}{3c} x^3 \tag{4.169}$$

At the singular point where $x = -1$, we have that

$$\tau = -\frac{2a}{3c} \tag{4.170}$$

which is clearly a finite value. Since τ is the time measured by the observer who approaches the event horizon, it follows that his clock will not record anything unusual at this point.

4.3 Notes

1. It is interesting to note in this context that an observer standing on the surface of the earth would *not* be inertial, since the spring in his meter would clearly stretch due to the weight of mass m_1. In that respect Einstein's view is very different from the Galilean one, which assumes that a stationary observer must be inertial.

2. The integration is performed with the added requirement that $x = 0$ when $\tau = 0$. This basically means that the observer reaches the center of the black hole at $\tau = 0$, and that all previous events are assigned negative times. There is nothing "unphysical" about such a choice, since time is always recorded with respect to a freely chosen reference point. In this particular case, we simply chose this reference point to coincide with the *final* event, rather than the initial one.

Further Reading

Arthur Eddington, *Space, Time and Gravitation*, Cambridge University Press, 1987.

Hermann Weyl, *Space, Time, Matter*, Kessinger Publishing, 2007.

David Bohm, *Special Theory of Relativity*, Routledge, 2006.

Steven Weinberg, *Cosmology*, Oxford University Press, 2008.

James Hartle, *Gravity: An Introduction to Einstein's General Relativity*, Benjamin Cummings, 2003.

Paul Dirac, *General Theory of Relativity*, Princeton University Press, 1996.

Stephen Hawking, *A Brief History of Time*, Bantam Books, 1998.

Alan Guth, *The Inflationary Universe*, Basic Books, 1997.

Kip Thorne, *Black Holes and Time Warps: Einstein's Outrageous Legacy*, W. W.Norton & Co., 1995.

S. Chandrasekhar, *The Mathematical Theory of Black Holes*, Oxford University Press, 1998.

Edwin Taylor and John Wheeler, *Exploring Black Holes: Introduction to General Relativity*, Addison Wesley, 2000.

Hermann Bondi, *Relativity and Common Sense*, Dover, 1980.

Stephen Hawking and Roger Penrose, *The Nature of Space and Time*, Princeton University Press, 1996.

Index

CPSIA information can be obtained
at www.ICGtesting.com
Printed in the USA
FSOW02n0658200916
25198FS

9 781609 274948